The Doctorate Blueprint

The Doctorate Blueprint

A Comprehensive Practical Guide for Graduates and Young Academics in Science and Engineering

ALESSIO MALIZIA

University of Pisa, Italy, and Molde University College, Norway

OXFORD
UNIVERSITY PRESS

OXFORD
UNIVERSITY PRESS

Great Clarendon Street, Oxford, OX2 6DP, United Kingdom

Oxford University Press is a department of the University of Oxford.
It furthers the University's objective of excellence in research, scholarship,
and education by publishing worldwide. Oxford is a registered trade mark of
Oxford University Press in the UK and in certain other countries.

Published in the United States of America by Oxford University Press
198 Madison Avenue, New York, NY 10016, United States of America

British Library Cataloguing in Publication Data

Data available

Library of Congress Control Number: 2025936022

ISBN 9780198927136
ISBN 9780198927129 (pbk.)

DOI: 10.1093/9780198927167.001.0001

Printed and bound by
CPI Group (UK) Ltd, Croydon, CR0 4YY

The manufacturer's authorised representative in the EU for product safety is
Oxford University Press España S.A., Parque Empresarial San Fernando de Henares,
Avenida de Castilla, 2 – 28830 Madrid (www.oup.es/en or product.safety@oup.com).
OUP España S.A. also acts as importer into Spain of products made by the manufacturer.

Links to third party websites are provided by Oxford in good faith and
for information only. Oxford disclaims any responsibility for the materials
contained in any third party website referenced in this work.

*To my beloved wife, Claudia, my soulmate, and my lovely daughter,
Sofia, for her brightness and gentleness.*

*In memory of Francesco Matera, a caring uncle, an extraordinary executive,
and a mentor whom we all miss dearly.*

Acknowledgments

Writing this book has been a profound journey that would not have been possible without the support and encouragement of many individuals.

I am thankful to the various scholars and authors whose remarks and suggestions laid the foundation for this book. Their advice has not only informed my writing but has also deepened my appreciation for their involvement in this process. I am grateful to Prof. Steven Adams and Prof. Susan Grey for involving me in the Research Degree Board at the University of Hertfordshire. I learnt a lot from both. I feel privileged to be a member of the board of the national PhD in AI and Society at the University of Pisa and would like to thank Prof. Dino Pedreschi for inviting me to join the board. I am particularly grateful to Prof. Vincenzo Ambriola, whose breadth and depth of suggestions have been invaluable to me. I am particularly grateful to Prof. Alan Dix, a friend and mentor, whose notes on research have provided me with much inspiration. I want to extend my sincere thanks and appreciation to Professors Paloma Diaz and Ignacio Aedo (Universidad Carlos III de Madrid) for their mentorship during my first PhD students' supervisions, and I would also like to acknowledge Professors Stefano Chessa and Massimo Pappalardo (University of Pisa) for the inspirational discussions we had around the coffee machine.

This book could never have been written without the experience gained and the lessons learnt whilst supervising PhD students in different institutions. I am particularly grateful to Professors Teresa Onorati and Andrea Bellucci (Universidad Carlos III de Madrid) for being my first PhD students and, in a sense, being guinea pigs for this book. I am extremely thankful to Dr Tommaso Turchi for stimulating discussions on all topics related to PhD studies, and with whom I discussed the draft of this book on a whiteboard at the old Saint John's building at Brunel University London many years ago. I would also like to thank all PhD Students of the first cycle of the National PhD in AI & Society at the University of Pisa, particularly Marta Doveri, Isacco Beretta, Martina Cinquini, Lorenzo Mannocci, and Serena Versino.

All friends and researchers extraordinaire of Genziana et al. (a collaborative circle) for stimulating discussions: Dr Andrea Beretta, Dr Luca Pappalardo and particularly Dr Filippo Chiarello for helping me scout research methods in engineering.

Finally, to everyone who contributed in ways both big and small, I extend my heartfelt thanks.

Contents

Introduction

Many people characterise obtaining a doctorate as a path that is emotionally exhausting, academically challenging, and occasionally even existentially confusing. Technical proficiency, perseverance, and self-motivation are as important as scientific curiosity on this journey into the unknown. Even though a PhD is the highest level of academic instruction, few people go into this stage with a clear plan. Unstructured expectations, the burden of original research, and the drive to publish overwhelm many students. On the other hand, young supervisors who are thrown into the simultaneous roles of manager and mentor frequently find it challenging to strike a balance between leading research and mentoring younger researchers. In response to these difficulties, this book provides an organised, experience-based, and practical manual for negotiating the PhD and the initial years of establishing an academic career.

In contrast to conventional academic guides, *The Doctorate Blueprint* demystifies the hidden curriculum of PhD research, including time management, scientific writing, networking, and coping with setbacks. It does more than just provide formal degree requirements. It offers useful information about multidisciplinary cooperation, successful publishing techniques, and supervisor-student dynamics. This book is a practical guide to academic research, whether you are a young faculty member looking to improve your mentoring style, an ambitious PhD candidate looking to expedite your research process, or a current student experiencing mid-journey difficulties. It gives you the skills you need to not only survive but also thrive in the rapidly changing fields of science and engineering through practical examples, tried-and-true tactics, and direct guidance.

I have been supervising several PhD students from different backgrounds, mainly in computer science, engineering, and digital design. All my students were interested in things related to technology from a scientific and engineering perspective; nevertheless, this book can be used by various students from different backgrounds who are interested in technology. I have been teaching over the years in computer science or computer engineering departments, as well as in a school of creative arts, where it was impressive to learn the number of digital tools digital artists use. I have been working with designers, mainly in digital or industrial design.

The Doctorate Blueprint. Alessio Malizia, Oxford University Press. © Alessio Malizia (2025).
DOI: 10.1093/9780198927167.003.0001

I have condensed all my expertise in supervising students in different areas but with a common interest in user-centred research and technology. I have learned a lot from the students. We have learned a lot together; indeed, one of my favourite metaphors for doctorate studies is a journey that the supervisor and the student take to investigate a research question. As it happens in every journey, there are ups and downs, but the outcome is mutual growth. A student starting a PhD is enthusiastic but a bit lost and usually feels overwhelmed by the number of things to do, tools to use, and methods to learn; that is precisely the purpose of this book: to offer help through a systematic approach to completing a PhD. This book will be helpful for students as well as for supervisors. Supervising or mentoring a post-graduate student working on a PhD is quite challenging, and hopefully, this book will offer the tools to manage such an exciting endeavour systematically.

Completing a PhD has at least two sides: the student and the supervisor, and their relationship. Sometimes, a supervisory team might be involved, and we will discuss this too. In this book, boxes can be found throughout the chapters; specific advice on mentoring during the different phases of a PhD study is offered in the boxes for supervisors. Such boxes will also be helpful to read for students: they suggest some valuable approaches to the supervisor based on my prior experience or can be used by the students themselves to improve communication with the supervisor.

This book starts with discussing what research is to help students reflect on why they like research and want to embark on a PhD study. We will focus on some critical questions: what is a doctorate? Why is it worth doing a PhD, and how do you break away from the undergraduate mentality? We will discuss some strategic guidance for students and supervisors, plus how to schedule for completion.

The first step to doing a PhD is crafting an appropriate research question; as they say, your research is as good as the research question. Spending time learning the language applicable to the research questions is crucial, as is learning, with the supervisor's help, to break the research question down into sub-research questions. The time required to complete a PhD is limited, so 'don't let the perfect be the enemy of the good' (Voltaire); a PhD candidate can spend an entire life investigating a research question. Still, a PhD lasts a limited period; therefore, under a supervisor's guidance, a student has to decide which part of a research question is worth digging into during the PhD studies and leave the rest for the future. You can never wholly investigate a good research question. If you can, it might not be that interesting.

At the same time, perfection can also be an excuse for inaction. People tend to overestimate the time and effort required to produce results, avoiding tackling an exciting research question that might appear too challenging. A PhD candidate should try not to be overwhelmed by challenges; an old saying says, 'How do you eat an elephant? One bite at a time' (Desmond Tutu).

To properly formulate a research question, it is essential to have a good understanding of the state of the art. A good literature review is crucial to understanding the challenges around particular topics of interest and where there is space for novelty and impact. Writing an excellent literature review takes work; many aspects deserve consideration, including skills to be developed by the student. Selecting the appropriate sources, reading academic literature, and understanding the relevance of research outputs are all critical skills for writing a good literature review. How to structure and write an excellent literature review is another crucial aspect to learn once adequate references are collected.

Successively, efforts are normally dedicated to finding an appropriate framework for the research. A good framework is inspired by an adequate research methodology that will guide all aspects of research activities performed during the PhD study. Different research methods in science and engineering will be explored, including a general discussion of research strategies and theoretical vs. conceptual frameworks.

Once an initial research question is formulated based on gaps and challenges identified in the state of the art, together with an appropriate methodology to develop the research, the ideation and co-creation phase can be used to generate a set of initial ideas and hypotheses. Diverse brainstorming techniques will lead to early prototypes and requirement gathering. Rapid prototyping is employed to quickly develop mock-ups to test the hypothesis and ideas formulated in the previous phase.

Prototype development is an iterative process that continually refines the ideas and functionalities of prototypes to test different aspects of the research question. Refining the prototypes involves evaluating both qualitative and quantitative aspects; this book will describe many different qualitative and quantitative techniques.

The dissertation section is dedicated to one of the focal tasks of the PhD study: writing a PhD dissertation. In this section, we will discuss planning, structuring, and writing techniques. A PhD study includes not only a dissertation but also, in different forms, a viva to pass to be awarded a PhD formally. The presentation skills section focuses on the viva and the skills needed to present the research at academic conferences. Indeed, dissemination is crucial in research to test the validity of result by peer review. Writing an academic paper is an important skill to learn. Still, we will also focus on writing for academic conferences and journals, and a series of tips on how to publish in top venues.

That leads to a series of tools to help students with the different phases just described: reference management systems for the literature review, mind maps for knowledge organisation, writing, planning, and data analysis software will all be part of the PhD toolkit, together with different research development frameworks, including tools and skills that a PhD student is expected to master during the studies. A comprehensive guide on artificial intelligence tools that can be used in all the different phases of doctoral studies will also be provided, whilst suggested readings will complement the student's research culture, plus some more specific books on writing, research methods, and experiments. Those are all collated in a section on inspirational and valuable books.

The book concludes with a 'what next' section that describes early career developments and consolidating one's own research lab through fundraising and grant management. Throughout the book, boxes are dedicated to suggestions for supervisors, with relevant advice in every section and phase of the PhD journey. The advice is based on my experience and will complement the information given to students well, providing sound guidance and a starting point for academics when supervising PhD students.

1

Embarking on the doctoral journey

What is research?

Before applying for PhD studies, the first question that should be considered is: 'Is research something I enjoy?' Reflecting on this aspect is a valuable initial step before committing to a period of rigorous work spanning three to six years, depending on the country chosen for pursuing a PhD. Prospective students are often advised to take a moment to contemplate the nature of research before making a decision.

However, that is not an easy question to answer since there are several types of research; in general, what we mean by 'research' in academia is a research project to generate new knowledge in a specific community. A research project is when someone experiments to find the answers to a research question or to solve a problem.

Experiments come in various forms—ranging from the classic image of a scientist in a lab coat working with vials and beakers to conceptual thought experiments reminiscent of Einstein at a chalkboard.

One characteristic of academic research is its systematic approach based on a formal scientific method. It investigates research questions guided by existing theories used to formulate hypotheses. These hypotheses can be rejected or supported by the experiments.

Understanding research and whether it is appealing is a crucial question to bear in mind before an application for a PhD is submitted and before a prospective supervisor examines a research proposal.

The 'Research techniques' notes by Alan Dix (2004) are a good starting point for learning more about research, offering some basic materials and definitions that will turn out to be very useful in different phases of your PhD. In particular, it gives a brief overview of nearly everything involved in working on a research project in a compact and accessible way. The notes were written for final-year projects, so the assumption is that there is no particular background in the research. The first part of the notes focuses on research, listing three topics: 'What is research, should you be doing it, and how do you do it?' My main suggestion to students here is to take a good look at it as an overview of all the phases involved in a research project and to focus on the 'What is research?' question at this stage.

The Doctorate Blueprint. Alessio Malizia, Oxford University Press. © Alessio Malizia (2025).
DOI: 10.1093/9780198927167.003.0002

In the notes, there are two good definitions of research with two slightly different perspectives: the scientist in a lab coat paradigm or the British Library historian who investigates all sorts of books and references to compile and collate state of the art.

Furthermore, two other categories of researchers are mentioned in the notes: the social scientist and the journalist. As a prospective PhD student, various roles might be assumed and interchanged, aligning with one or more of the four categories based on the stage of your studies: initially, a state-of-the-art compilation might be prepared (historian); subsequently, hypotheses and research questions could be developed (scientist/investigative journalist); experiments may be designed from these (social scientist/lab coat scientist), and qualitative or quantitative data might be analysed to confirm or reject the hypotheses.

This brief introduction to research should give an idea of what to expect and help prospective students make an informed choice, but I recommend a reading list to my students. These are books that might give future PhD candidates a broad overview of research and stimulate their curiosity or dissuade them from starting a research project; I will not recommend reading them all, but picking at least one or two that feel more suitable to the aspects of research a candidate might like most: ideas and invention, scientific thinking and revolutions, how we know the things we know, and so on.

Bill Bryson's *A short history of nearly everything* (2004) is an excellent introduction, including how we went from the Bing Bang to reaching the Moon, passing by the Renaissance, and the Industrial Revolution. It is a book about reflecting on how we know what we know, including discoveries in chemistry, physics, and geology, accessible to a broad audience. For example, questions like what we know about the Earth's centre or what a black hole is are all tackled in this book with an entertaining style.

Guardian columnist Ben Goldacre is famous for debunking fake science. In his book *Bad science*, Goldacre (2010) discredits big pharma, mistreaters of evidence, and mass media for promoting 'the public misunderstanding of science'. One of the debunking episodes is about a vitamin pill manufacturer, Matthias Rath, who claimed his vitamins were the 'natural answer to AIDS', a reminder that ignorance of the truth can be deadly. Throughout the book, the scientific method, specifically hypothesis testing, will be applied to validate or reject hypotheses and support research questions. However, a recollection of all the classic mistakes to avoid when attempting to infer the truth about a phenomenon from data analysis will be especially emphasised.

Thomas Kuhn's *The structure of scientific revolutions* (2012), first published in 1962, revolutionised the understanding of how scientific breakthroughs happen. In his work, Kuhn shows how paradigm shifts in science do not emerge from gradual experimentation and discovery but occur outside of what is.

Ideas: A history from facts to Freud (2006) is a very inspiring book that tells the history of ideas from the dawn of time to modern times. Peter Watson analyses how the earliest ideas might have arisen from original beliefs such as Greek myths and ancient gods to crusades, the Renaissance, Gutenberg, and the invention of print, to mention a few. Studying the process and history of ideas might spark prospective students' curiosity about innovation and inspire them to create new knowledge. It's crucial in a PhD dissertation to show how you're adding to this new knowledge in a specific research area.

Furthermore, a group of four books suggested by Prof. Miguel Nacenta for students to read in their free time provide valuable insights to reflect on at the beginning of a PhD. I would suggest looking for those who might help prospective students reflect on whether they like research and want to embark on such a journey. The books are *The craft of research* (Booth, Colomb, & Williams) and *The elements of style* (Strunk & White), *A PhD is not enough!* (Feibelman), and *Getting things done* (Allen).

A very inspiring talk given by Richard Hamming, 'You and Your Research'[1], in 1995, focused on the observations and research he conducted on why only a few scientists make significant contributions. Hamming delved into the properties of individual scientists, including their abilities, traits, working habits, attitudes, and philosophy. He emphasised the importance of the creative spark coming from within oneself, the significance of smart people finding interesting angles in results, and the process of building a substantial story from a small, solid experimental result. Hamming's talk aimed to provide insights into the factors that contribute to scientific success and the mindset required for impactful research endeavours. In my opinion, there are particularly three pieces of advice that PhD students should focus on:

1. The best approach to research is to work on the right problem at the right time and in the right way, and nothing else. That is, focus on what is important and worth full attention.
2. Working environment, that is, the open-door versus the closed-door office. Keep our minds open, explore with others, and learn what is important by cross-contaminating ideas. While a closed door will help focus, we might end up working on a specific thing that is not as important as a contribution to a field; by keeping the door open, we can learn about new problems and ideas that can stimulate our interest and also get an insightful critique of our approach to a problem.
3. Take time to stop and think about the important problems in your subject of study. Try not to drown in everyday work; think about what's relevant and see if you can attack one of those problems. A good strategy can be to set

---

[1] https://www.youtube.com/watch?v=a1zDuOPkMSw.

aside some time, for instance, 10% of your time on Friday afternoons, not Mondays, when you can easily be drawn back to urgent things.

Finally, I would suggest looking at the 'Business Model Canvas'. Just like a beginner learning to play chess might be instructed to start learning the endgames, I recommend looking at this framework. For someone interested in a research career, guidance will be provided on different aspects of developing skills and succeeding in academia. Before commencing the path, it is crucial for goals to be understood and established. Clear goals can prevent the temptation to stray towards time-consuming yet unrewarding activities. Analogous to the chess metaphor, devising a successful plan to checkmate an opponent becomes difficult without a clear understanding of the desired achievement.

The 'Business Model Canvas' is part of a collection of materials that will help PhD dissertation or academic book authors (Medium, 2015). It consists of an analytic framework developed for start-up firms but works well for researchers and academics at different stages of their careers, including PhDs; it helps to reflect on the actions before committing to a particular stream of research. The framework introduces nine items to consider when planning your research career, from reflecting on the novelty and impact of your future research (value added) and considering the communities that are going to be impacted by your research (user segments and relationship with users) to international markers of esteem (citations and academic recognition) and key partners. While some parts of the framework will be empty at the beginning of your research, all those items will start filling up as your career progresses and will provide good guidance to set up a plan, usually starting with your PhD thesis.

What is a doctorate?

According to a blog post[2] published by Franklin University: 'the doctorate is the most advanced degree you can earn, symbolising that you have mastered a specific area of study or field of profession. The degree requires a significant level of research and articulation. Those who earn the degree must have researched a subject or topic thoroughly, conducted new research and analysis, and provided a new interpretation or solution into the field.' It means conducting a research project, as mentioned in the previous section. Therefore, it is essential to make sure a candidate likes research or at least has a general understanding of what research is before starting a doctorate.

Prospective students might have heard doctorate or PhD degrees used as synonyms, which is usually correct. According to *findaphd.com*, a doctorate is a

[2] https://www.franklin.edu/blog/what-is-a-doctorate-degree.

qualification that awards a doctoral degree and earns you the title 'Doctor'. The PhD is the most common type of doctorate. However, other doctorates, such as professional doctorates, are usually awarded in specific areas involving more practical or professional projects (e.g. in the arts or in design). Essentially, all PhDs are doctorates, but not all doctorates are PhDs, known as 'Doctor of Philosophy', meaning someone mastering an area of study.

Box 1.1

To clarify, I suggest showing students two diagrams:
The fish diagram and the circle diagram. The fish diagram, part of a set of online materials called 'The art of doing a PhD', represents the whole process of completing a PhD in phases that, in abstract form, take the shape of a fish,* where the first third of the research project is open and broad, including literature exploration and mining the state of the art to formulate a research question. In contrast, the second part, consisting of a substantial half of the PhD work, is more focused or narrow, summarising the literature, building a system, devising an experiment, and writing papers to conclude with the last bit that opens up again with the write-up phase and identification of future works and challenges.

The circle diagram† by Matt Might, Professor of Computer Science at the University of Utah, clarifies how you contribute to knowledge creation through your research project by starting with a circle representing all human knowledge and showing a student's contribution as a small dent pushing the boundaries of the state of the art. It is small but relevant since the contribution will increase the knowledge in a specific community. Often, a student will start overenthusiastic, imagining to cause a paradigm shift in a research area with their work. However, in most cases, it will be the small dent in the circle diagram, which is a more realistic contribution achievable in a specific amount of time, for example 3–6 years of an average PhD degree. The contribution will be relevant independently of the size of the dent, especially for the community addressed in the research project.

* https://www.bardram.net/the-art-of-doing-a-phd/.
† https://www.openculture.com/2017/06/the-illustrated-guide-to-a-phd-12-simple-pictures-that-will-put-the-daunting-degree-into-perspective.html.

Philip Guo, Professor of Cognitive Science at UC San Diego, mentions hardships students may face when studying for a doctorate in his blog[3] on advice for

[3] https://cacm.acm.org/blogcacm/ph-d-s-from-the-facultys-perspective/.

early stage PhD students (2012). The student typically starts with great potential; after all, the student successfully applied and was accepted into a PhD programme. Unfortunately, quite a few students fail to achieve their full potential in research achievements. Why? It is not because they fall short on creativity, intellectual abilities or scientific skills; instead, other elements contribute to the shortfall pinpointed by Guo as a lack of resilience, perseverance, metacognition,[4] and self-discipline. The remedy, Guo suggests, is self-reflection and mentorship. In particular, students have to face three challenges, as Guo mentions: uncertainty, isolation, and project scoping. Uncertainty and isolation—the feeling of not knowing whether you are progressing and the lack of understanding of what PhD work entails from people surrounding you—can be counteracted by feedback from a mentor (regular meetings with your PhD supervisor should be attempted) and from an external community of experts by submitting papers and getting peer reviews. Due to the open-ended nature of a PhD, the project scope entails questions such as: How much work is required for an acceptable prototype or experiment? How much is enough to submit a paper? And for a completed dissertation? (Guo, Nov. 2013). All those questions are part of this book, and generally, a clear answer cannot be obtained, but it will come with experience and mentorship from the supervisor. It is important not to feel overwhelmed by trying to answer those questions at once but to seek help internally from a supervisory team or mentor and externally from communities working on the same topic, that is, attending a doctoral consortium at an international conference.

The relationship with the supervisor is crucial for the success of your PhD and is neglected in many of the books advising on doctorate studies.

The significance of the relationship with the supervisor regarding the success of a PhD journey cannot be overstated. This is a crucial point, yet it is frequently missed in books that provide advice on PhD studies. The relationship that develops between a doctoral candidate and their advisor is essential in guiding the research process, offering guidance, and impacting the doctoral experience in general. This relationship includes mutual understanding, good communication, and the creation of a cooperative and encouraging atmosphere in addition to academic assistance. Acknowledging and maintaining this relationship is essential to overcoming obstacles, creating a supportive research atmosphere, and ultimately helping a PhD programme to be completed successfully.

In 'The supervisor-student relationship: The problem of conflicting 'mixed metaphors'[5] Avison et al. (2013) describe how particular supervisor-student relationships can be seen as one of several metaphors, for example, a journey, a marriage, apprenticeship, and servitude.

---

[4] https://cft.vanderbilt.edu/guides-sub-pages/metacognition/.
[5] http://elibrary.aisnet.org/Default.aspx?url=https://aisel.aisnet.org/cgi/viewcontent.cgi?article=1033&context=bled2013.

> **Box 1.2**
>
> ---
>
> In my experience, those categories are spot on, and those types of relations repeat over time. In particular, some relations with their known problems affect younger or early career professors who might dedicate more time to their students, often establishing some unhealthy dependency from the student where the boundaries between the roles are crossed, initially with enthusiasm but ending up in disappointment over time. I have always established a journey where I, as a mentor and the student, will embark on the research around a new topic together, facing open-ended challenges where the mentor can suggest specific approaches. Still, the student is encouraged to try their ideas and solutions. In such a way, it is mutual growth, and everyone benefits from it. Many institutions have recently established a process in which there are continuous checks on the PhD student's progress, typically paced around 8–12 months. In my experience, detecting any problem or struggle at an early stage is vital to preventing detrimental relations between students and supervisors from being discovered later on. Different remedies are available at an early stage, from offering extra support to the student or the supervisor to involving a more expert colleague in the supervisory team or, in extreme cases, identifying another supervisor. It is also essential for the student to understand which kind of relations has been established with the supervisor and seek advice to remedy the situation before it can affect the progression of the PhD studies.

What should a PhD student look for in a thesis advisor? The supervisor's research, productivity, and qualities as a doctoral advisor are likely the most important things to look for. The career stage is also another good point to consider. Usually, an early> to mid-career supervisor will be less experienced but might have more time to dedicate to students than a well-established full professor running a lab with tens of PhD students. A well-established professor running a lab with many PhD students will have less time, but all will be organised and managed. While you might feel like a cog in the machine, you would also benefit from a lab where you could access more experienced students.

On the contrary, an early career advisor might spend more time investigating new approaches with their students but might be more chaotic, busy fulfilling tenure requirements, and have fewer students as part of a lab. Clearly, both situations have pros and cons and depend on the student's level of autonomy and self-discipline. Overall, the main question you should consider when choosing your PhD advisor is whether you will be successful if you work with this professor. Most students wonder if the professors are friendly, if they are tenured, and if they have a good sense of humour, and so on. That matters only if you can complete your PhD.

Box 1.3

Meeting minutes are ideal for improving communication between students and supervisors and detecting issues early. For instance, I would always ask my students to take notes during meetings and send them back to me by email, many times in a short form, that is, bullet points. This way, the supervisor can be confident that the student correctly understood what was advised and double-check communication issues; it's a powerful tool to help plan the deliverables agreed upon during meetings, such as dissertation chapters, literature reviews or papers. Currently, there is software to manage postgraduate students and supervisors (e.g. Research Supervision and Management System—RSMS), such as automatically checking students' progress and all the phases that come with a PhD from the initial registration to the final exam (e.g. viva). These will automatically remind of meetings and offer standard report tools to write meeting minutes. Improving communication is also essential to quickly establish the student and supervisor's expectations; this will also clarify the level of independent study required from a PhD student. A questionnaire to measure expectations can be used to explain what is expected from the student and supervisor. Open questions or a Likert scale can be used to inspect those expectations; here are some questions one might ask: Is the supervisor responsible for selecting a topic, or must students be responsible for setting their own topic? For instance, use a Likert scale of 1–5, where 1–2 means the responsibility to choose the topic will fall into the supervisor's hand, 4–5 is for the student to choose their own topic, or three is balanced between the supervisor and the student. Other questions might be around the different phases of a PhD: Should the supervisor develop an appropriate programme of study and timetable for research, or should the supervisor leave the development of the programme of study to the student? Should the supervisor assist in the write-up if necessary, or should the write-up only ever be the student's own work?

Supervisors and students come in all sizes and shapes and organise their work with different approaches. Here are a few tips taken from Prof. Mick Watson's article 'Tips for PhD students and early-career researchers' (2016, 27th Oct), published on his blog and further refined by me, which might be very useful.

- Take the lead: this is your research project. Remember, a PhD is not a taught course, so students must learn to investigate their ideas and plan and conduct experiments to test them. Don't be afraid to explore.
- Read: plan carefully and reserve time for reading; as I always say: 'no input, no output'. To get ideas, experiment, and write, you need to read, and

typically, there is extensive literature, so try to organise yourself and use some reference management system to catalogue your readings (e.g. Zotero).

- Write: learn how to write efficiently; being a researcher is also being a writer, so many activities involve writing a dissertation or a paper, writing reports, proposals, emails, and so on. If you prefer to avoid writing, you should consider whether you want to start a career in research.

- Engage: try to engage with people, and don't be afraid to ask questions. Research is a social activity; go to meetings, seminars, conferences, and workshops and try to be active by asking questions and engaging people with interesting ideas in conversations. Particular conference sessions and symposiums are dedicated to PhD students, such as a Graduate or Doctoral Consortium. Furthermore, a great opportunity is to serve as a student volunteer, where you can help organise a conference in exchange for waived fees or accommodation and connect with other students at a similar stage who might become part of your future academic network.

- Record: try to keep a journal or a notebook, physical or digital, recording everything you do. From lab notebooks to wikis or personal blogs, which work well in my experience, make sure to keep track of all your activities and thoughts.

- Twitter and LinkedIn: those social networks are valuable tools in a research toolbox. Twitter allows you to get in touch with thousands of researchers worldwide, and you can follow discussions and news on specific subtopics using hashtags. It is also an excellent chance to contribute and have your say in the community of interest, plus many announcements regarding conferences and journals are usually advertised on Twitter. LinkedIn is not just for job hunting. Students will find a lot of special interest groups, can connect to scholars, write collaborative articles, and attend conferences and meet-ups.

- Learn: try to constantly progress and learn new skills, especially those in demand, and focus your research using those skills. Your supervisor can help you with essential skills such as writing a proper literature review, designing experiments, and evaluating outputs. Many universities offer doctorate college courses on how to do research, especially in the first year of your PhD. Beyond basic training, try to find which skills and techniques are in demand and learn them.

- Plan: in my experience, you would rarely achieve anything if you didn't have a plan. Considering a doctorate will take a few years, planning is crucial. Different techniques are available, but it is vital to have a plan from day one and stick to a realistic goal. Try to include all the phases, from literature review to hypothesis and research question formulations, from prototypical design and development to testing and experimentation, including the write-up.

- Speak: research is about communicating results to the rest of the world or at least to communities of experts interested in your topic. Try to interact with people often, give talks at conferences or in your own department, attend meetings, and don't be afraid to have your say. Submit abstracts and give talks whenever possible; talking about your research will help you externalise your thoughts, and collecting feedback will help refine your research objectives.
- Realise: your job is really about pushing the boundaries of human knowledge. It's hard, challenging every day, and can be stressful, but it brings unexpected rewards and contributions to society. Remember, you have a supervisory team and the doctorate college infrastructure surrounding you to support your work and encourage your completion.

Prof. Garrett Johnson (Questrom School of Business, Boston University) has put together guidance for PhD students that includes similar tips. Still, he complements it by mentioning various pieces of advice.

The guidance is split into three main sections: production, marketing, and writing.

In production, idea generation is described using a '3D filter' and questions such as where do ideas come from? Where do ideas go, and what do I do with my ideas? These are answered briefly but effectively; for instance, turning a '3-line ideas -> 1 page -> 5 pages -> Paper!' He also highlights that writing is thinking, and by writing down ideas, PhD students reflect on and refine them. Another piece of advice in the same session is to be careful with time management ('time is your enemy') and to chart a course and move towards it by, for instance, 'Hypothesis-driven problem solving, i.e., pick a good direction and move forward'. The final advice in this section is about talking to others about your own work: 'classmates, advisors (regularly), faculty members, guest speakers'; also be organised, for instance, using 'bibliographic software (e.g. BibDesk, Mendeley, Zotero) to organise all your papers and all citations'.

In the marketing sections, Prof. Johnson reminds students that production and marketing are ideal for a career in academia, but what is intended by marketing in this case? Caring about communication and developing communication skills by 'practising writing, speaking, presenting and seeking training, feedback, & practice opportunities'. Learn that 'good academic writing is clear and concise' and 'good presentations are rehearsed'. There is a section of this book on writing where we will examine in depth how to develop good academic writing skills, for example, by following a 'sense of style' and learning about presentation skills for conferences, meetings and meetups. Finally, marketing at the academic level is intended to publish long and short papers at conferences or in journals, write reports or abstracts, and write for social networks such as Twitter or LinkedIn.

Box 1.4

A vital component of higher education is doctoral supervision, which mentors prospective researchers through the demanding process of study and discovery. Its fundamental ethical obligation transcends beyond advice given in schools (Lee & Bongaardt, 2021). It entails protecting candidates' mental and emotional health, promoting intellectual development, maintaining justice, and honouring diversity.

The underlying principles that govern the mentorship and interaction between supervisors and candidates form the basis of ethics in PhD supervision. Integrity and trust serve as cornerstones, requiring sincerity, openness, and moral behaviour. Respect for autonomy allows candidates to be recognised as autonomous scholars and gives them the freedom to decide for themselves while getting advice. Fairness and equity emphasise the importance of preventing prejudice, creating an inclusive atmosphere, and offering equal chances and resources. Professional behaviour requires respecting moral principles, keeping a proper distance, and using prudence while handling private material.

Supervisors are responsible for a variety of tasks. The first and foremost responsibility is academic guidance, which includes providing helpful criticism and scholarly guidance and fostering an atmosphere encouraging intellectual inquiry. Mentoring goes beyond the classroom to support applicants' goals for both their personal and professional growth. Support for emotional and mental health recognises the strains associated with PhD research, encouraging supervisors to pay attention and provide assistance when required. Supervisors are required under ethical research conduct to ensure that their candidates follow ethical guidelines.

Managing the complexities of PhD supervision requires resolving several issues. The supervisor-candidate relationship's inherent power dynamics necessitate deliberate attempts to reduce imbalances through honest communication, unambiguous expectations, and readily available support. When it comes to handling conflicts that arise from miscommunication or differing expectations, conflict resolution techniques become crucial. A difficulty in fair resource allocation is to ensure equal opportunity for all candidates through transparent techniques and equitable standards for evaluation.

If success is to be achieved, some guidance is needed to facilitate the candidate's development. While this guidance will be provided by their supervisory team and the resources established by the doctorate college at the university they have chosen, having their own plan for professional development is advisable.

Box 1.5

I suggest students consider researcher development frameworks such as the Vitae Research Development Framework* (Vitae RDF). A framework will help you set your goals for the doctorate and future career as a researcher/academic. The Vitae RDF covers a range of disciplines and assists with identifying the PhD students' strengths and prioritising development goals. It includes a broad series of area descriptors and phases to cover from their early-stage PhD to mature researcher. It is handy for understanding where to focus students' efforts.

* https://www.vitae.ac.uk/researchers-professional-development/about-the-vitae-researcher-development-framework.

It is pretty easy to be distracted by various activities that could harm candidates' PhD work over time. Practical activities like committing to personal projects, assisting fellow students, and writing that one-pager that was requested can all contribute. However, if not appropriately prioritised, these activities could be detrimental to students' dissertations. In such cases, a research development framework like Vitae could assist.

My experience supervising PhD students has brought me to give some advice on what to think about when students are considering a PhD study and how to write a doctoral proposal for their admission:

1. To prepare a doctoral proposal, I always suggest that students download a PhD dissertation from the Internet and look at it to understand what kind of contribution is expected. Depending on the area they want to work in, their supervisor or some fellow students can pass on a previous dissertation for them to examine. I advise students to browse articles on *phdcomics.com*. This website includes articles and comic strips on 'the life or lack thereof in academia', mainly focused on PhD students and academic supervisors. It gives some humorous advice and a sense of what students can experience during their PhD life.

2. Starting with broad scopes, most doctoral proposals are typically broken down into subtopics, and the ones of most significant interest are then selected. Given that the next few years will be dedicated to working on these topics, they must be truly embraced. The boundaries of those subtopics of interest should be clearly defined.

3. Identify the most relevant scholars in the specific area of interest within the last five to 10 years; ensuring that the research remains up to date is essential. Consider which materials produced by those scholars stand out and the reasons behind their significance. Traverse the research outputs

generated by those scholars in reverse chronological order, and it might be discovered that, around 20 years ago, a seminal book relevant to the interests was produced by the same scholar. Measures of 'quality' for the sources, such as the scholars' h-index—a metric indicating productivity and impact in terms of citations of their publications—or the scientific reputations of the venues (scientific journals, conference proceedings, books, etc.) are to be determined.

4. Students must keep an annotated bibliography of primary and secondary references and use a reference management system (e.g. Zotero or End-Note) to sort those out. A record of the metadata associated with every book, paper, or source, including authors, titles, publishers, and so on, should be maintained. Additionally, notes should be kept, often referred to as an annotated bibliography, detailing why each source is relevant to the research. One limitation observed in many bibliographies is the need to appreciate the sources critically. Students fall into the mistake of reporting a huge list representing all that has been written about a particular field. In contrast, a critical evaluation of the sources will require motivation to determine why a reference has been cited and which parts are relevant to their research. Candidates must evaluate those in a summary form (literature review or state of the art) and elaborate on the most important, compelling issues that need addressing.

5. From the literature review, consideration should be given to the line of thinking, that is, methodology, that may be required to contribute to the field. Following such a methodology, expectations for an original contribution to the area should be outlined. Whether introducing a new line of thinking or offering a different perspective on an established one, the anticipated contribution should be clearly defined.

6. Testing ideas against the state of the art and assessing their effectiveness is crucial. Identifying a gap in the current state of the art is important. The theoretical modelling of the field may have needed to be improved. A specific issue with the ongoing theory might have been overlooked. A gap in the discipline should be considered. These clues play a helpful role in shaping the original research question and conceptualising the hypothesis to be tested during the doctorate.

The research work plan summarises critical literature and works, including frameworks, prototypes, and systems, that form the foundation of the research background. Emphasis is placed on the relevance of this background to the field, acknowledging potential strengths and shortcomings in the review. Research questions are crafted by utilising gaps and limitations identified in the state of the art. The potential contributions of the work to the field of study are described while setting a broad scope for the research. It is acknowledged that the initial scope will evolve and be refined throughout the PhD studies, alleviating the need for detailed specifications from the outset.

A final piece of advice involves framing the research within the context of the school, college, department, and so on, where the PhD studies will be conducted. Consideration should be given to how the research aligns with the institution's plan—identifying individuals with the relevant expertise who could guide the refining of the research project. Available labs that seamlessly integrate with the research should be explored. Reflection on how the institution and fellow PhD students might benefit from the work is recommended. This self-inquiry helps position the research within the broader context of the host institution, fostering collaboration and mutual benefit.

At this point, the primary question that could be considered is whether a PhD is what students want. The initial step involves sitting down and contemplating the future career, assessing the necessity of pursuing a doctorate. Consideration should be given to the desire for a career in academic or industrial research. Academic job opportunities are generally limited, and including the tenure track necessary for achieving a professorship entails a prolonged period before substantial financial stability is reached. Except for top research centres, a master's degree is likely the primary requirement for key positions in the private industry. It holds true, however, that as a doctorate holder, expertise in the field becomes a potential advantage in the hiring process of an industrial lab.

During PhD studies, the mastery of research execution and effective communication of results in publications and research outputs will be acquired. Research validation and source and output reliability assessment will involve engagement with key stakeholders within academic communities or companies. Undertaking a PhD necessitates several years of rigorous work, be it on a full- or part-time basis. Employers may perceive a PhD holder as too expensive or overqualified, although this perception hinges on the specific requirements of the position under consideration. For example, a PhD in academia has the potential to lead to a full professorship, possibly even encompassing roles such as President or Chancellor. Private companies increasingly acknowledge the potential of PhDs, particularly in science, engineering, and technology domains, where innovation plays a crucial role in maintaining competitiveness.

Doctoral exams differ significantly between schools and countries. The most common methods involve a faculty committee questioning the candidates during a traditional oral examination about their research plan, expertise in the field, and other relevant subjects. It might also involve a more thorough analysis of the field or the dissertation proposal defence. Some colleges will take published papers, a portfolio of work, or sizeable research projects in lieu of or in addition to the candidacy exam. Alternatively, candidates could demonstrate their knowledge through published works rather than taking tests. Some schools less frequently assess candidates on the basis of a substantial project or dissertation in lieu of oral exams. The main factors used to assess a candidate's suitability for a PhD are the project's calibre, scope, and originality. Lastly, some universities or countries have abandoned conventional PhD examinations in favour of evaluating candidates primarily based on their research accomplishments, courses, and dissertations.

The duration for the completion of a PhD programme can vary significantly, depending on several variables such as the programme, the country, the country's educational system, and the topic of study. After completing a bachelor's degree, a PhD can be obtained in four to eight years of full-time study on average. It may take longer in some fields because of the nature of the research and dissertation requirements, such as science or engineering. Time variation is also influenced by factors such as experiments, data collection and analysis, study difficulty, and data analysis. The PhD programme typically entails research, coursework, comprehensive exams, and completing a doctoral dissertation. Some programmes, however, might offer expedited pathways based on prior coursework or research expertise, or they can have different structural options.

For example, in the United States, obtaining a PhD usually requires multiple steps, which adds to the lengthier time than in certain other nations (e.g. the UK or Europe); before pursuing a PhD, many students in the United States seek a master's degree. Although some programmes offer a direct entry into a combined master's and PhD track, which might decrease the total timetable, this step can take about two years. The PhD programme typically takes four to six years to finish once a master's degree is earned or after enrolling in a combined programme. The subject of study, research requirements, courses, dissertation preparation, and teaching or other obligations affect how long it takes. To develop subject-matter knowledge, the first one to two years typically consist of coursework, seminars, and exams. Conducting original research, gathering and analysing data, and writing the dissertation take up most of the time. This stage may need two to four years or longer, contingent upon the complex nature of the investigation. Many PhD candidates additionally take on teaching or research assistantship responsibilities, which can lengthen the programme's duration but offer invaluable experience.

Nonetheless, a few alternatives and quick programmes are available; for example, certain universities allow outstanding applicants or those with substantial research experience to enter PhD programmes directly, skipping the master's degree stage. These programmes have the potential to shorten the total time.

In order to shorten the time needed for completion, several fields or multidisciplinary programmes may provide expedited pathways. These programmes may require prior experience or coursework and frequently involve larger workloads.

Financial aid, fellowships, or available scholarships can also affect the length. Students who receive full financing may be able to concentrate solely on their academics without juggling employment obligations.

On the other hand, because of less emphasis on coursework and a more focused research approach from the start, PhD programmes in some countries, such as the UK and portions of Europe, can be shorter. The lengths of these variations are determined mainly by each country's academic traditions, programme designs, and educational systems. Compared to the USA, PhD programmes in the UK and other parts of Europe frequently take a more concentrated and efficient approach. Three to four years is the average length of a PhD programme in the UK and Europe. This shorter duration results from these programmes' increased emphasis on research, which places more emphasis on the dissertation and less on coursework.

Students in these programmes typically get into their research projects right away, delving deeply into their selected subject matter. Compared to US institutions, their education is less organised, which lets students focus more on their research promptly. The thesis or dissertation serves as the programme's focal point. Most of a candidate's time is devoted to writing their dissertation, gathering information, interpreting findings, and performing original research. The programme's shortened duration is largely due to this concentration on autonomous research. A selected supervisor or advisory group provides doctoral candidates with close supervision and guidance. This mentorship is essential to guaranteeing the research project's advancement and calibre.

Although the coursework is less intensive, certain programmes might nevertheless offer seminars or instruction in academic writing, research techniques, and other pertinent skills. These are usually less comprehensive than those in American programmes, though. PhD students frequently receive financing or stipends in various European nations. They are able to fully concentrate on their studies and research thanks to this financial support, which eliminates the need for them to take on more jobs. Candidates present their thesis to an examining committee after finishing their dissertation. A PhD degree can only be obtained by passing this oral defence.

Other countries, for example, countries in Asia or Oceania, share some similarities with the approaches mentioned earlier, and more specifically, the average length of a PhD programme in China is three to five years. Coursework, extensive examinations, and a strong focus on research and dissertation writing are all part of them. Similarly, three to five years in Japan are often required for PhD programmes there. With an emphasis on original research, they entail coursework, research, and producing a thesis or dissertation. In contrast, in India, PhD programmes vary greatly, usually lasting between three and seven years. The length of time varies according to the topic of study, the need for research, and the advancement of the individual. While in South Korea, four to six years is the average length of a PhD programme. They entail research, coursework, and dissertation preparation, with a growing focus on globalisation and collaborative research.

Full-time PhD programmes in Australia typically last three to four years. They include coursework at the start of the degree and a research project that ends in a thesis. Similarly, three to four years are typically required for New Zealand PhD programmes. With occasional deviations based on the area and institution, they strongly emphasise research and thesis writing.

There are, nevertheless, some common elements; for instance, completing a thesis or dissertation and conducting original research are priorities for PhD programmes in Asia and Oceania. While some programmes start out with some coursework or training, the focus soon switches to independent research. Candidates frequently collaborate closely with an advisory committee or supervisor throughout their studies. STEM professions frequently have lengthier durations because of experimental activity and data collection. However, duration and structure can vary greatly across disciplines.

It is evident that PhD programmes differ greatly also depending on the topic or area of study. Because of the variety of research methods, approaches, and scholarly traditions that exist within each discipline, there can be significant differences in the requirements, structure, and length of time. For instance, humanities and science/engineering PhD programmes are very different in terms of focus, approaches, and prerequisites, as shown in Table 1.1.

Although both routes result in a doctorate, the precise focus, approaches, and conclusions differ greatly depending on the area of study selected.

In the context of this book, our primary focus will centre on PhD programmes within the domains of science and engineering. Science and engineering PhD programmes differ globally because of variances in educational systems, research priorities, programme designs, and academic prerequisites.

Depending on the institution's strengths and the demands of the sector as a whole, programmes may have a regional focus. Certain nations may have specialised in particular engineering domains, such as renewable energy or automotive engineering. Certain technology or research fields, including artificial intelligence (AI), cybersecurity, or biotechnology, may be given priority in some locations over others.

Doctorate programmes can differ greatly in length. PhD programmes in science and engineering typically last four to six years in the United States and other countries, though they may take shorter (for instance, three years in some European countries) or longer. While some programmes focus heavily on research from the start, others have demanding academic requirements that students must complete before starting their research.

Funding sources and availability for doctoral candidates can vary. For instance, within the UK, funding for postgraduate studies usually consists of tuition fees and a stipend to cover living expenses, unlike in many other countries where PhD students are frequently hired as teaching or research assistants with stipends. Self-funded doctoral candidates do so without receiving outside funds from research councils, universities, or other funding organisations. These students use loans, personal savings, family support, part-time employment, or other sources to pay for their living expenses and tuition. Some countries' programmes may have

Table 1.1 Differences between the humanities and engineering PhD programmes

PhD in humanities	PhD in science/engineering
The study of human culture, history, language, literature, philosophy, the arts, and social sciences is frequently the focus of doctoral programmes in the humanities. The emphasis is on analysis, interpretation, and critical thinking. It often entails a lot of reading, writing, and research in disciplines like anthropology, history, literature, and so on.	More technical and focused on research and innovation, PhD programmes in science and engineering are offered in disciplines including computer science, electrical engineering, mechanical engineering, and so on. These courses strongly emphasise problem-solving, science, experimentation, and technology development.
Qualitative research methodologies, textual analysis, historical research, and cultural artefact interpretation are frequently employed in humanities studies. A thesis or dissertation that adds fresh perspectives or interpretations to the body of current knowledge is usually the outcome of the research.	Experimentation, modelling, analysis, and quantitative methods play a major role in many research domains. Developing new technologies, algorithms, and systems or resolving challenging technological issues are common project tasks. In the fields of engineering and science, doctoral dissertations may contain theoretical contributions, software systems, algorithms, or prototypes.
Humanities PhD programmes could demand extensive examinations, seminars, and the finishing of a sizable dissertation built on original research.	Coursework, qualifying tests, laboratory or field research, and the writing of a dissertation demonstrating original research contributions in the subject are frequently required for these programmes.
Humanities PhD programme graduates frequently go on to work in publishing, writing, cultural organisations, government agencies, or academia as professors or researchers.	Graduates of these schools frequently find employment prospects in research and development positions, working on cutting-edge technologies or improving current systems in industry, academia, government, or tech startups.

closer linkages to industry, providing greater chances for internships, business partnerships, or post-graduation employment placements.

Some nations' programmes may strongly emphasise cross-border cooperation, team research initiatives, or industrial relationships that give students access to special opportunities.

While some nations may conduct their PhD programmes in science and engineering exclusively in English to draw in international students, others may provide these programmes largely in their local tongue. Programmes in some nations may be more prominent or globally recognised than others because of the standing of the respective universities or the general calibre of the research output.

The composition, structure, and expectations of the panels involved in the evaluation process, which includes the thesis defence, may vary. Expert assessors are gathered in PhD defence panels for science and engineering to appraise a candidate's research and academic path. Usually consisting of eminent academics, subject matter specialists, and occasionally outside examiners, the panel represents a variety of viewpoints in the candidate's field of study. The composition often consists of committee members, the candidate's principal advisor or supervisor, and academics honoured for their contributions to the particular field of study. This assembly guarantees a thorough assessment examining the breadth, originality, and academic value of the candidate's work.

In science and engineering PhD defence panels, the candidate often gives an oral presentation outlining their study approach, findings, and conclusions. A thorough questioning session where panellists delve into the nuances of the research typically follows this presentation. Inquiries can go deep into topics such as experimental designs, theoretical underpinnings, data interpretation, and broader research implications. Candidates could also be asked about the limitations of their work and potential directions for further research. Consequently, the defence procedure necessitates a deep comprehension of the study conducted and the capacity to communicate its importance clearly.

During these defences, PhD applicants face particularly high expectations, which include providing a thorough display of their competence, critical thinking skills, and intellectual contributions. In addition to demonstrating expertise in their field of study, candidates should be able to defend their methods clearly and convincingly, explain the significance of their results in the context of higher education, and justify their conclusions. In addition, applicants must demonstrate the ability to participate in intellectual dialogue by confidently and shrewdly responding to panel questions. In addition to marking the end of the candidate's PhD journey, the defence attests to their preparedness to contribute significantly to their academic field.

Comprehending these distinctions can be vital for candidates contemplating doctoral programmes in Engineering and Computer Science globally since it facilitates the synchronisation of individual research inclinations, professional aspirations, and preferences with the provisions of disparate establishments and programmes.

Summary

This chapter serves as a comprehensive guide for both prospective PhD students and early career supervisors. It begins by assisting readers in deciding whether to pursue a PhD, offering insights into the commitment and implications of doctoral studies. The text then provides a concise overview of research fundamentals, including various methodologies and the systematic, scientific approach required in academic research.

For early career supervisors, the chapter offers practical advice on setting clear student expectations and evaluating PhD proposals effectively. It aims to help candidates understand the realities of doctoral studies, emphasising the importance of transitioning from an undergraduate mentality to a more independent, research-focused mindset. The chapter also addresses potential challenges that PhD students may face.

Establishing a positive and productive relationship between supervisor and student is another key focus, with the chapter exploring strategies for effective communication and collaboration. The discussion concludes with an extensive examination of doctoral examination processes, highlighting the significant variations between different institutions and countries. By covering these diverse aspects, the chapter provides a realistic and comprehensive view of the PhD experience, equipping both students and supervisors with valuable knowledge to navigate the challenges and opportunities of doctoral studies.

2

Research methods

Conducting a literature review

Organising a systematic literature review

The *Chambers 20th Century Dictionary* defines research as *a systematic investigation towards increasing the sum of knowledge.* The first step towards achieving a systematic investigation is a systematic literature review.

It translates to reading what has been written by others, encompassing academic papers, books, conference proceedings, articles, and so on. Additionally, consideration must be given to interviews and opinion articles in magazines and social media groups. Furthermore, examination might be required of what has been made by others in terms of prototypes, open-source software, artefacts, and so on.

The 'Research techniques' notes by Alan Dix (2004) serve as a useful starting point for planning a literature review. As recommended by Dix, the first step involves locating literature through keyword searches. Various academic search engines, ranging from Google Scholar to Scopus or ResearchGate, can be utilised; nevertheless, searches will be conducted based on keywords. After reading to understand the vocabulary, meanings, and terminology used in a certain field, the keyword search is then refined.

A few articles or books will probably be found after a preliminary keyword search, prompting the desire to expand the search. Dix recommends expanding the collection of sources by moving forward or backwards in time. One approach involves moving backwards in time by exploring the bibliography of a paper. It is possible to come across a citation from a previous book that might be regarded as seminal or to come across an older work by the same authors that might be relevant. The deeper you delve into the bibliography, the more likely you are to uncover progressively older articles. For example, a book from 2010 might cite a relevant paper from 1997, which in turn references an article from 1985, and so forth. Another strategy involves moving forward in time, and Dix recommends utilising a citation index in this scenario. Information about which other papers and books reference the one you are aware of can be obtained. Google Scholar provides an easily accessible means of acquiring this information by conducting a search for the article you are presently reading. The search results will include a

The Doctorate Blueprint. Alessio Malizia, Oxford University Press. © Alessio Malizia (2025).
DOI: 10.1093/9780198927167.003.0003

list of all the sources, including papers, books, and proceedings, that reference the specific article.

Through a keyword search, a multitude of references will be accumulated. The subsequent step involves employing a filtering strategy to identify the most pertinent sources. Initially, the title, keywords, and abstract can be utilised as a basic filtering criterion. An alternative approach involves employing the citation count as a criterion for filtering relevant publications. If a book has been cited numerous times, it is likely to be pertinent to the topic. While this is considered a somewhat crude measure, citation counts can be found on platforms such as Google Scholar or Scopus. Additionally, on websites like ResearchGate, the number of reads can also be found, providing another heuristic to be considered for selecting relevant sources.

Once a source, such as a paper, has been filtered, I would suggest to begin by skim-reading and then extracting information for the reading records. Metadata extracted from the source, including title, keywords, and abstract, should be recorded. Additionally, one should incorporate personal keywords or tags supported by many reference management systems (e.g. Mendeley).

Alan Dix (2004), in the 'Research techniques' notes, in a section entitled 'Other people's work—what they make', describes how to unpack the implicit experience embedded in other people's artefacts, frameworks, software, or systems.

Neglecting the mistakes made by others in constructing prototypes within the same research area may lead to the repetition of those errors. Learning from the work of others is crucial. Examine prototypes, frameworks, systems, software, and artefacts developed within the same topics chosen for your research project. Incorporate reviews of these outputs into your literature review. For every output deemed relevant to your research, Dix suggests examining 'What is good and Why is it good', as well as considering 'What is bad and Why is it bad' and 'Why do it this way?' These questions contribute to the critical appraisal of existing outputs and generate ideas for enhancing artefacts. For example, combining ideas from various systems can lead to the development of an entirely new system. Alternatively, reusing an existing system but adapting it to a different context, such as modifying an interface for the blind, is another approach. Additionally, improving and rectifying mistakes evident in current systems are a valuable avenue for research.

Writing a systematic literature review

A good piece of advice about how to write a systematic literature review comes from Tay's (2020) article on Nature, 'How to write a superb literature review'. The article mentions that while literature reviews are essential resources for scientists, few researchers are skilled at writing literature reviews. A good literature review should 'provide historical context for a field' and identify its future trajectory up to open challenges that can inspire further research.

The advice in this article comes from different researchers and scholars, providing a set of principles:

- *Be focused and avoid jargon.* Writing a literature review can improve understanding of the history of a field and identify open challenges that can trigger ideas for future contributions. Prof. Wentin Zhao mentions that 'a common problem for students writing their first reviews is being overly ambitious'. Zhao advises students to consider the literature review as different from a textbook and offer a more focused discussion. Consider writing the review with the awareness that the audience may include non-experts in the field, such as other PhD students investigating the literature for the first time. Zhao continues: 'A good review should also avoid jargon and explain the basic concepts for someone new to the field.'

- *Have a process and develop your style.* Prof. Bozhi Tian's advice is to ask yourself, 'What is the purpose of this review?' Usually, it is to get a chance to 'contribute to the insights to the scientific community', identify opportunities for contributing to the knowledge, and introduce some novelty by mapping the existing state of the art to your research topics. After clarifying the purpose of the review, it is good to discuss the outline with your supervisor. Successively go through multiple iterations to ensure that all the relevant sources are included and don't overlap with the existing state-of-the-art reviews. Tian also reminds students to picture themselves as 'artists of science' and encourages them to develop their writing and presentation skills.

- *Timeliness and figures make a huge difference.* Ankita Anirban, *Nature Review Physics* editor, advises, 'It is not enough for a review to be a summary of the latest growth in the literature: the most interesting reviews instead provide a discussion about disagreements in the field.' Contradictory results and contexts offer an opportunity to properly landscape topics within a field. As highlighted by Ankita, such critical appraisal is crucial to recognise broad trends and identify trade-offs or underlying mechanisms that could reconcile the various conclusions in the literature. She concludes by pointing out the role of figures, which, in my opinion, is crucial, together with tables to make sense of and summarise a literature review. In her own words: 'An important role of a review paper is to introduce researchers to a field. For this, schematic figures can be useful to illustrate the science being discussed in much the same way the first slide of a talk should.'

- *Stay updated and be open to suggestions.* It takes a long time to complete a literature review, generally a year or more. Sometimes, when completed, it might look obsolete since other relevant papers and books are published.

It might later be discovered that a certain topic should have been mentioned and, as a result, must be included in the review. Staying updated means revising your literature review at least every six months during your PhD study and completing it in iterations. Prof. Yoojin Choi recalls:

> During this time (up to 2-year timespan to complete a literature review, revision and publication), many new papers and even competing reviews were published. To provide the most up-to-date and original review, I had to stay abreast of the literature. In my case, I made use of Google Scholar, which I set to send me daily updates of relevant literature based on keywords.

- *Make good use of technology.* Paula Martinez-Gonzales, a PhD student at Cambridge University, suggests skilling up on software and trying to automate some daunting tasks by extracting all relevant literature related to a specific topic. She recounts:

> As I was researching the review, I noticed a trend in which some papers were consistently being cited by many other papers in the field. . . . That was when I decided to code a small application to make my literature research more efficient. . . . The code then identifies the relevant papers and creates a citation graph of all the references cited in the results of the search. The software highlights papers that have many citation relationships with other papers in the search and could, therefore, be called seminal papers.

  She used PubMed, but other websites such as Google Scholar or Scopus would offer the same type of metadata about publications.
- *Reference Management software and Collaborative Writing Tools.* The article concludes with a table summarising some of the most popular reference management software: EndNote, Mendeley, Zotero, and Paperpile, categorised by cost, storage capacity in the Cloud, and support for collaborative working, among others; Collaborative writing tools—Manubot, Overleaf, and Google Docs—are classified by price, availability of mobile devices, writing language, and so on. This table is handy. Indeed, it is crucial to create a toolbox of software that might come in handy while working on a doctorate study. This book has a chapter dedicated to assembling a PhD toolkit.

Box 2.1

Advise students to try to provide a critical appraisal of sources and not a list of 'here is all that is known on this topic'.

I always suggest that my students identify papers in two categories: champions and borderlines. Champions will form the core of your literature review (seminal books, highly cited papers matching the main focus of their dissertations, frequently accessed articles, etc.). Borderlines will be included in a successive iteration, expanding the review to an interesting but somewhat different point of view that might complement the landscaping of the field.

Finally, I found that Pinker's (2015) book, *The sense of style*, provides good guidance on improving and caring about students' writing styles. Good communication is essential to research.

Structuring a systematic literature review

Considering that a systematic literature review can take a long time and include a vast amount of material, a crucial question is how to structure it. A well-structured literature review should consist of the following:

1. Evidence before the study: In this section, a description of all evidence considered before the commencement of the study is provided, stating:
   a. The sources that were searched, including databases, journal or book reference lists, and so on, are detailed.
   b. The criteria utilised to include or exclude studies, encompassing the precise start and end dates of the search, are outlined. This is not limited to English-language publications.
   c. The search terms employed in the process are specified.
   d. The quality (risk of bias) of the evidence considered is discussed.
2. If applicable, a pooled estimate derived from a meta-analysis of the evidence is presented.

To conduct a meta-analysis of the sources and consequently compile a systematic literature review, it is imperative to establish the parameters of the research and delineate the methods employed to identify pertinent papers. The search terms, including their plurals, employed are explicitly specified based on the relevant topics.

A defined timeframe, indicating the start and end dates of the search, is established.

The scope of the search is clarified, indicating whether the search was confined to titles and abstracts to streamline the manual filtering process for irrelevant papers. The websites, such as Google Scholar, and digital libraries, for example, including ACM DL, IEEE Xplore, DBLP, and so on, utilised for the search, are enumerated.

The contributions presented in the collected sources, encompassing papers, books, and so on, can be categorised into a taxonomy based on the nature of their approach. This taxonomy aids in classifying the entire body of literature and facilitates a detailed analysis of the various approaches proposed within each category. Some potential categories include:

1. *Theoretical approach*: papers that primarily focus on developing and presenting theoretical frameworks, models, or conceptual paradigms without immediate practical implementation.
2. *Pragmatic approach*: works that emphasise practical applications, real-world scenarios, or hands-on methodologies, with a focus on practical outcomes and solutions.
3. *Technological approach*: contributions that centre around the development, implementation, or evaluation of technological solutions, tools, or systems.
4. *Empirical research*: papers reporting on empirical studies, experiments, surveys, or data-driven investigations that contribute to the understanding of a particular phenomenon.
5. *Review and synthesis*: literature reviews, systematic reviews, or synthesis papers that consolidate and analyse existing knowledge in a specific area.
6. *Case studies*: documents that present in-depth examinations of specific cases, scenarios, or instances to draw insights or lessons.
7. *Methodological contributions*: works that introduce novel research methodologies, experimental designs, or data analysis techniques.
8. *Policy and regulatory analysis*: papers focusing on the examination of policies, regulations, or legal aspects relevant to the subject matter.
9. *Cross-disciplinary perspectives*: contributions that integrate insights from multiple disciplines or provide a cross-disciplinary perspective on a particular topic.
10. *Educational contributions*: materials that contribute to educational aspects such as textbooks, instructional guides, or pedagogical approaches.

Assigning papers to these categories will allow for a systematic analysis of the diverse approaches presented in the literature, providing a structured understanding of the field.

The meta-analysis can be enhanced through the incorporation of both quantitative and qualitative analyses of the sources. In the quantitative analysis, the number of papers by publication year, with reference to the short-listed sources, can be utilised. For example, if it is observed that certain years witnessed heightened popularity in the topics of interest, a corresponding increase in the number of papers can be noted. Additionally, a graph can be constructed, encompassing papers by publication year and type (journal, conference proceedings, books, etc.). Potential effects can be discovered, such as the shift towards a prevalence of journal publications over conference proceedings after a certain period. This may suggest the maturation of the topic into a more established field, where substantial studies, typically published in journals, gain preference over rapid communications found in conference proceedings. Another aspect involves the creation of a table incorporating sources, especially papers, which enumerates authors publishing more than one paper in the literature review. This facilitates the identification of individuals and research groups engaged in these topics.

Qualitative analysis commences with the establishment of a citation graph pertaining to the sources examined in the review. Subsequently, the citation graph can be linked to the topic of the papers, mapping them within the initially created taxonomy. This enriched citation graph offers an overall depiction, facilitating a comprehensive understanding of the field of interest. Moreover, approaches (theoretical, technological, etc.) can be mapped to leading research groups in the area, along with a subset of specific papers that potentially represent a particular approach to the field.

A standard approach to writing a meta-review has been suggested by the PRISMA group (2009). The PRISMA statement consists of a checklist[1] and a flow diagram.[2] It is intended to be accompanied by the PRISMA explanation and elaboration document. It is a methodology used mainly in the medical field but rigorous enough to be inspirational for many other areas to write systematic literature reviews and meta-analyses. By meta-analysis, we intend to review and compare existing studies and their results. According to the PRISMA group,

> a systematic review is a review of a clearly formulated question that uses systematic and explicit methods to identify, select, and critically appraise relevant research and to collect and analyse data from the studies that are included in the review. Statistical methods (meta-analysis) may or may not be used to analyse and summarise the results of the included studies. Meta-analysis refers to the use of statistical techniques in a systematic review to integrate the results of included studies.

For example, the checklist employed in the PRISMA method is highly organised. It enumerates the elements that one may consider incorporating into systematic reviews, including but not limited to title, abstract, introduction, methods (for filtering and analysing relevant sources), results (such as the synthesis of findings), discussion (providing a summary of evidence), and funding.

A report written by Barbara Kitchenham,[3] entitled 'Procedures for performing systematic reviews', July 2004, offers good guidance for structuring a systematic literature review.

---

[1] https://www.prisma-statement.org/prisma-2020-checklist.
[2] https://www.prisma-statement.org/prisma-2020-flow-diagram.
[3] https://www.inf.ufsc.br/~aldo.vw/kitchenham.pdf.

Although the advice in the report was created by combining three already-used guidelines for medical researchers, the guidance works well in general. The guideline covers the three stages of a systematic review—planning, carrying out, and reporting. Notably, structuring a meta-analysis is not considered, but this book is focused on advice for writing a literature review at the doctorate level.

A systematic review entails a number of distinct tasks. The number and sequence of tasks are suggested in various ways by current systematic review guidelines.

The report divides a systematic review's stages into three basic stages: organising the review, carrying it out, and reporting it.

The stages associated with planning the review are two: *identification of the need for a review* and *development of a review protocol*. The stages associated with conducting the review are five: *identification of research, selection of primary studies, study quality assessment, data extraction and monitoring,* and *data synthesis*.

Reporting the review is considered a single-stage phase. The report presents each phase in detail and offers a good summary of all the activities required for a proper literature review.

Reading an academic paper

Following the structuring of the systematic literature review, the subsequent step involves reading the sources, taking notes, and completing mini-abstracts along with all relevant information that is intended to be recorded within the chosen reference management software.

Engaging with an academic paper differs from reading other forms of literature, as advised by Prof. Mark Pallen in 'How to read a scientific paper'. The recommended approach involves examining the paper's organisation, typically following a general structure known as IMRAD (introduction, methods, results, and discussion). Additionally, attention should be given to elements such as the title, abstract, authors, acknowledgements, declarations, references, tables, figures, and legends. It's important to note that the process of reading an academic paper doesn't necessarily need to adhere to a linear order, contrasting with reading a novel.

A more effective strategy involves prioritising elements such as the title, abstract, figures, tables, introduction, results, and discussion, with a consideration of the methods section coming later. Specifically, it may be possible to comprehend tables and figures independently of the surrounding paragraphs initially. Subsequently, reading those paragraphs might provide a deeper understanding of the presented data. Reading an academic paper requires an active approach, as Prof. Pallen suggests, encouraging the use of highlighters, underlining text, making comments or posing questions, and taking notes. Prof. Pallen emphasises the importance of persistence, advising readers to 'read and re-read, spiralling in on central points' when faced with initial difficulty in comprehension. Adopting a 'question-asking mode' is considered crucial, given that, despite publication, papers may contain undetected faults or issues overlooked by reviewers. Engaging in the active dissection of an academic paper allows for the exercise of peer-reviewing skills, ensuring the validity of all claims made within the paper. Prof. Pallen suggests a scheme to evaluate a paper that, in my opinion, is very effective (to use as well as for guidance for peer-review activities):

- What questions does the paper address?
- What are the main conclusions of the paper?
- What evidence supports those conclusions?
- Does the data actually support the conclusions?
- What is the quality of the evidence?
- Why are the conclusions important?

An excellent framework to analyse a paper is offered in 'How to read (and understand) a social science journal article' (Laubepin, 2013). In 'Anatomy of a journal article' section, all the parts of a typical journal article are dissected and described in a table, including the elements (title, abstract, introduction, etc.), what they are, and what they tell you. For instance:

- *Element:* Title
- *What it is:* the title presents a concise statement of the theoretical issues investigated.
- *What it tells you:* what this paper is about.

The second part of the article describes different strategies for reading an academic paper depending on your purpose.

The initial step involves identifying the specific reasons for your interest in reading the paper and determining the information or insights you aim to extract from it. Laubepin invites us to ask: 'Are you interested in the author's theoretical perspective? Her findings? Her methods? Her data? Are you interested in getting a sense of the research that has been done on a specific topic/issue? Knowing the answer to these questions will determine your reading strategy'.

The following section introduces a table designed to assist in the formulation of a reading strategy based on one's purpose. For instance, the table is divided into components: the sections of a journal article (title, abstract, introduction, etc.) and the purpose (what is being sought). For example, understanding the overarching concept in the paper to capture its essence or directing attention to specific details.

The final sections are about the difference between structural reading and close reading. Structural reading is a form of close reading but applied to the structure of the source, for example a book, where one reads through the title, introduction, table of contents, and blurb to get a perspective on it; it works for journal papers too. It's like a 'mental scaffolding', according to Laubepin: 'It helps the reader to determine whether she wants to spend time reading the text and how closely she wants to read it.' Close reading, instead, is a more detailed and focused reading where 'you may want to stop at every paragraph to summarise what is being said, reflect on the arguments being made, and evaluate the quality of the evidence being presented'. Laubepin advises: 'Take ownership of what you read: mark the text up, jot down questions, comments or observations in the margins, highlight important passages/quotes, and take notes as you go. Interacting with the text in these ways makes it more likely that you will remember the information as well.'

Richmond Alake summarised advice from Andrew Ng, a renowned machine learning expert, in 'How you should read research papers'[4].

1. The steps suggested in Alake's article are: first, assemble a collection of resources, articles, books, and blogs on the topic of your research.

2. Track your understanding of every resource by using a table with sources plotted against your level of knowledge in percentage intervals. For instance, for every source, design the table to have column intervals representing your level of understanding of the corresponding source: 10–20%, 20–40%, 40–60%, 60–80%, and 80–100%. Reading at least 10–20% of the sources listed in the table is ideal for ensuring you understand how relevant that particular source is before digging deeper into it. For papers you consider more appropriate, it makes more sense to achieve a higher percentage and for some to fully understand the source. You might ask what a reasonable number of papers to read is.

3. Take structured notes; try to capture key discoveries, findings, and techniques presented in a paper; and condense them in mini-abstracts—something between 100–250 words—summarising the relevant things for your research from that particular source.

4. Alake suggests that you should 'be prepared to go through a paper at least three times to have a good understanding of its content'.
   a. 'The first pass consists of reading the title, abstract and figures.'
   b. 'The second pass consists of reading the following sections: introduction, conclusion, another pass through figures, and scan through the rest of the content.'
   c. 'The third pass of the paper involves reading all the sections within the paper but skipping any complicated maths or technique formulations

---

[4] https://towardsdatascience.com/how-you-should-read-research-papers-according-to-andrew-ng-stanford-deep-learning-lectures-98ecbd3ccfb3.

that might be alien to you. During this pass, you can skip any terms and terminologies you do not understand or aren't familiar with.'

5. 'Those conducting in-depth research into a domain can take a few more passes. . . focused on an understanding of the maths, techniques, and unknown terminologies presented within the paper.'

In conclusion, the article by Alake incorporates a set of questions to be considered as a guide for assessing one's comprehension level of the sources. This can serve as an excellent template for summarising academic papers:

1. Describe what the authors of the paper aim to accomplish or perhaps did achieve.
2. Suppose a new approach/technique/method was introduced in a paper. What are the critical elements of the newly proposed approach?
3. What content within the paper is helpful to you?
4. What other references do you want to follow?

Box 2.2

From my experience, I would say that to get a good grasp of the subject, students might start their literature review with an initial set of around 30–50 papers and then increase the number of papers during further interactions. Something between 150–300 references is considered the norm for a PhD dissertation, but there is no specific amount; it depends on students' critical appraisal of sources and the subject (humanities might require more sources than engineering).

Figuring out how to extract relevant information from papers and how to provide a critical review of the sources are essential skills to develop to achieve an excellent systematic literature review. Still, there is another skill that is also necessary: note-taking. The extraction of relevant information from papers can be guided by the four points outlined in the preceding paragraph. Nevertheless, the systematic synthesis of the contents of all the papers, book chapters, and articles to be read raises the question of how this can be achieved.

Box 2.3

I suggest my students write a 100-word mini-abstract for each source they read. By writing mini-abstracts, they will summarise the paper's importance for their specific work/research question. Furthermore, when they go back to

Continued

the literature review, perhaps after a while, through those mini-abstracts, they will never forget what the paper was about without the need to skim through it again. Another benefit of the mini-abstract strategy is that students will have a substantial amount of words when they assemble, for instance, a set of 150 mini-abstracts of 100 words each. Therefore, the literature review chapter is basically already sketched and needs only some tweaking.

In terms of note-taking, everybody has a strategy. Still, I have found the Cornell note-taking system[5] useful (Pauk & Owens, 2010).

The organisation of the notes using the Cornell system consists of splitting a note-taking file or notebook page into three sections: cue, notes, and summary. Cornell notes are, in fact, helpful in taking notes in class, but they also work for offline sources or online materials.

In the notes section, lines or paragraphs representing the source's central message are documented. These often include sections previously underlined during the initial reading or schematic representations of relevant elements, such as figures or tables. The cue section is conventionally completed after note-taking and serves as a method for making sense of the notes. This involves articulating questions raised by the notes to aid in organising the material or inspiring further readings. The cue section may also include summarised questions or elements of the paper expressed through paradigms or similes. The summary section, reserved for the 100-word mini-abstract, encapsulates the essence of the source, succinctly conveying the key insights gleaned from that particular reference.

Selecting the right sources

The next big question about a systematic literature review is selecting the sources or, in other words, from where to pick the papers, books, and articles you want to explore in an organised literature review.

The first point of contact is evidently the supervisory team, which is made up of experts who may provide an initial list of sources or indications of where to locate them, serving as the starting point for the exploration. Secondly, three types of sources can be considered: academic, online, and the so-called grey literature. Academic literature can be accessed in libraries, both on campus and in public libraries, or in digital libraries like ACM DL, IEEE Xplore, SpringerLink, and so on. Online resources, such as social networks, may offer free versions of certain materials that would otherwise require a subscription in more traditional digital libraries such as ACM DL and IEEE Xplore.

[5] https://lsc.cornell.edu/how-to-study/taking-notes/cornell-note-taking-system/.

Online-accessible networks exist for the exchange and browsing of papers. Platforms like ResearchGate allow users to request papers from authors who can share publications either privately or publicly. Within these social networks, users have the option to follow authors, initiate contact, or connect with research labs online. Common social networks such as Twitter and LinkedIn also serve as channels for accessing articles, papers, or blogs managed by researchers. Grey literature encompasses outputs such as articles, technical reports, or white papers generated by organisations beyond conventional commercial or academic publishing and distribution channels. This includes entities such as non-governmental organisations, think tanks, or companies.

All those sources of materials will generate a considerable amount of resources, so the next question is how to filter the most relevant and trustworthy sources.

To filter out a list of relevant and trustworthy materials, I recommend referring to three parameters: the reputation of the venue where the work has been published, the reputation of the authors, and the number of citations.

Initially, once a paper or book has been identified, reference can be made to a measure of impact in the academic community. This information can be obtained through the advisory team, which can provide a specific set of measures. This involves examining the venue's impact factor and its classification in Scimago or in 'Journal citation report' on the Web of Science. Scimago allows for a quick understanding of the journal's categorisation in terms of quartiles and keywords, offering a rapid assessment of the source's relevance and reliability. Conferences, on the other hand, may have different classifications, such as the CORE[6] conference ranking. In the case of books, the publisher serves as a reliable indicator; well-known publishers in the academic world, including Springer, Cambridge University Press, Oxford University Press, MIT Press, or Routledge, are often associated with credibility.

The reputation of the author plays a crucial role in confirming the quality of the source and also in recognising scholars in specific fields to ensure that no other relevant paper or book by the same author is overlooked. The H-index serves as a measure of an author's productivity and impact in terms of citations, introduced by physicist Jorge Hirsch in 2005 at the University of California. Calculating the H-index involves determining the number of papers (H) cited at least H times. For example, if an author has an H-index of 25 according to Google Scholar, it signifies that the author has 25 papers cited at least 25 times by other research papers.

---

[6] https://portal.core.edu.au/conf-ranks/.

Google Scholar[7] or Scopus[8] offer quick access to the author's H-index. In general, the H-index on Scopus or other academic sites is more robust than Google Scholar, which can be more easily tricked and includes in the H-index calculation a wide variety of outlets whose academic integrity is non-verifiable.

Other platforms where authors' reputations can be assessed include DBLP,[9] where all authors' publications are listed and categorised by types, such as books, journals, or conference proceedings. DBLP functions as a time machine, enabling the exploration of an author's list of publications backwards and forward over the years. This allows checking for any relevant publications by an author prior to the current literature or updates from an author who published a seminal work in the past. Another tool is Publish or Perish,[10] a software platform used to evaluate the impact of someone's research on academic communities. It provides comprehensive bibliometric data that proves valuable for conducting a literature review (Harzing, 2023).

The ResearchGate[11] website gives an indication of a scholar's reputation and impact on a research area; it measures the scientific reputation based on how peers receive all of the scholar's research. Different from the H-index, Research-Gate advocates the adoption of research interest. This metric captures the interest in a researcher's work across the scientific community for those who seek to have a metric to rapidly analyse their own and others' profiles to comprehend their contributions to science (Yu et al., 2016).

The Web of Science Researcher ID provides lots of statistics and bibliometrics about scholars, as well as the Orcid Id,[12] which provides a unique identifier for researchers and relates to all their contributions in terms of academic outputs.

Finally, a scholar's home page may provide good information about a scholar's achievements and, therefore, is a good resource for estimating the reputation of such a scholar's contributions to the research community.

The number of citations is a crucial indicator of how relevant a publication is, whether it be a paper or an article. We can determine how much other works have been impacted by a paper by counting how many times it has been mentioned in other works. The average number of citations varies by research topic; thus, while certain fields have fewer citations and even a few in the tens are regarded as relevant and influential, other fields only regard a paper as relevant when it has a few hundreds or thousands of citations. When an output significantly outperforms

[7] https://scholar.google.com/.
[8] https://www.elsevier.com/en-gb/solutions/scopus.
[9] https://dblp.org/.
[10] https://harzing.com/resources/publish-or-perish.
[11] https://www.researchgate.net.
[12] https://orcid.org.

the typical amount of citations in a certain field, it is regarded as very influential or seminal.

The citation count of an article can be readily accessed on Google Scholar, allowing for an examination of the distribution of citations to determine the years when the work was most relevant. This analysis helps discern whether the paper is foundational, still frequently cited, and enduring over time. Scopus provides a more robust measure of academic citations, often offering a citation count smaller than Google Scholar but considered scientifically sound; Google Scholar, instead, can be more susceptible to manipulation through cunning approaches that involve nesting the same article citations through hierarchies of websites. Furthermore, Google Scholar has a tendency to include citations made by blogs, websites, and even grey literature.

However, not all citations are equal in value, and a paper's worth can change based on the reason it was cited. Unfortunately, there aren't many experimental measures that attempt to gauge the importance of citations individually rather than as a total. A work by Zhu et al. (2015) proposing a strategy to categorise the worth of individual citations is an intriguing example. Although the paper is a fascinating experiment, it's important to read the introduction carefully because it describes multiple criteria to gauge citation counts.

Indexing and recording references

Following the filtration of references (papers, articles, books, etc.) that align with your research interests, the subsequent step involves recording these references. Two critical questions need addressing: what to record from each reference and where. I propose adding the following elements to your records for each reference: key points, details, keywords or tags, and a mini-abstract. Key points consist of brief sentences summarising the relevant information from the reference. Details include figures, graphs, and tables, serving as pertinent artefacts from the reference, along with keywords or tags derived from those suggested by the authors or created by you for easy categorisation. Finally, compose a 100-word summary, a mini-abstract that succinctly summarises the key points in a discursive yet compact format, facilitating later reference to remind you of the importance of citing the source in your literature review.

Where do you keep track of all this material, including references, keywords, tags, key points, and mini-abstracts? A reference management system. Reference management systems are tools that students, academics, and researchers primarily use to organise sources and provide accurate citations. These tools can serve as an ongoing library of references and can be linked to Cloud services that provide storage space.

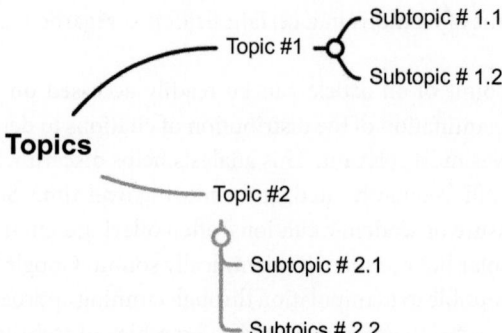

Figure 2.1 A mind map with topics and branches

Reference management software, including Mendeley, EndNote, and Zotero, come in a variety of versions; some are free, while others need a software licence. All of them, however, provide similar functionalities, particularly records to save data and metadata about references, as well as to generate bibliographies in various formats that students can then directly import into their dissertations.

Box 2.4

Mind mapping. At this stage, I recommend that my students use a mind map to organise and classify the ideas and key points collected from the literature review. Also, concept maps, which are less hierarchically structured and have a looser organisation, can be used for the same purpose.

A mind map is a visual organisation tool that typically takes the form of a tree-shaped diagram (see Figure 2.1). Usually, there is a group of ideas or essential elements that have been taken out of sources that can be related to one another, organised around a main idea, and linked with words, pictures, or hyperlinks. Big concepts are intimately tied to a central issue, but additional related ideas branch out from those major concepts to build a hierarchical diagram that resembles a tree. There are also Concept Maps, which are less hierarchically structured and have a looser organisation. 'Mind mapping' is a broad term that is often misunderstood. When people try mind mapping, they frequently discover that their maps are disorganised and do not aid understanding. Additionally, concept maps and mind maps are frequently used interchangeably, which makes it challenging to make efficient concept maps or mind maps. However, using mind mapping and concept mapping effectively can significantly enhance the organisation of your literature review or writing, aid in comprehension, clarify concepts, and save a significant amount of time. Some fundamental definitions:

- *A topic* is a distinct thought or note that might or might not be related to other topics.
- *A branch* is a group of concepts that branch off from the main concept in the centre of a mind map. Branches typically do not link to one another.
- The goal of a mind map or concept map is for each topic to reflect a complete, distinct thought. This is a piece of knowledge. *Knowledge objects* are not a collection of independent or tangentially linked concepts. It is up for interpretation as to what constitutes a 'full' thought.

A mind map is a categorised list of concepts and ideas. Around a core concept, ideas are categorised into branches. Associated with the fundamental theme are major topics (sections within your dissertation). Subtopics are related to these major topics. Branching results from them. Compared to the broad ideas of the main subjects, the subtopics are more focused. There is no connection between the branches. An excellent mind map should have a 'trunk' (the major idea), 'branches' (the important topics), and 'twigs', just like a tree (subtopic). Each subject on the map must be a distinct, stand-alone concept (knowledge object).

Vast amounts of knowledge can be organised and structured with a mind map, facilitating the planning of the literature review chapter and the categorisation of information by providing order. In conducting a systematic literature review, a significant amount of data must be organised in a logical manner. However, assistance in comprehending the significance or meaning of individual pieces of information is not provided by mind maps. Before incorporating data into the mind map, a comprehensive understanding of the information is necessary. If difficulties are encountered in understanding something, the creation of an idea map may be considered. The connections between various ideas are illustrated by a concept map.

For example, start with the centre node that represents your primary subject of interest or inquiry. The main themes or concepts linked to your research should be represented by branches extending out from the core node. Use sub-branches for further categorisation that can exist for each major theme.

Create sub-branches to symbolise each major theme's subthemes or subtopics in order to elaborate on each. These subthemes may be based on several facets, viewpoints, or fields of inquiry connected to your main subject.

Include nodes to reflect pertinent studies, authors, or particular works under each subtheme or subtopic. The main conclusions, approaches, or theoretical frameworks related to each study can be included.

To link relevant research, authors, or concepts in the mind map, use lines or arrows.

Thanks to this graphic portrayal, you can better comprehend how many literary works are related to one another and how they advance your learning as a whole.

Use a section or node on your mind map to draw attention to any gaps in the literature or areas that need more research. You may also add your own thoughts or possible study channels based on the body of existing research.

Create a new branch to describe the various methodological techniques used in the literature, if applicable to your research. Include nodes that indicate the numerous authors' individual research methodologies.

Finally, include a branch or sub-branch to indicate important theoretical frameworks or conceptual models that are common in the literature, if appropriate. Include nodes for certain theories or models and their links with the subject of your study.

Conversely, a concept map is open-ended. There isn't a main concept in it. Each subject on the map must be a distinct, stand-alone concept (knowledge object). A line connecting one topic to another defines the nature of the link between the two topics. You can use concept maps to visualise relationships and concepts that are challenging to understand. However, because concept maps do not arrange your thoughts in a logical, linear manner, which you will need to accomplish in order to write up your systematic review, they do not assist you in organising your literature review (see Figure 2.2).

It frequently happens that students attempt to create a mind map but end up with a concept map. This occurs when there are several lines without a hierarchy and the concepts are not properly organised into branches of primary subjects and subtopics (classification of ideas into main ideas and subtopics of the main idea).

Because ideas in concept maps are not presented in a logical, linear order, the resulting diagram does not assist in organising your systematic literature review. Additionally, concept maps generated by mind-mapping software lack

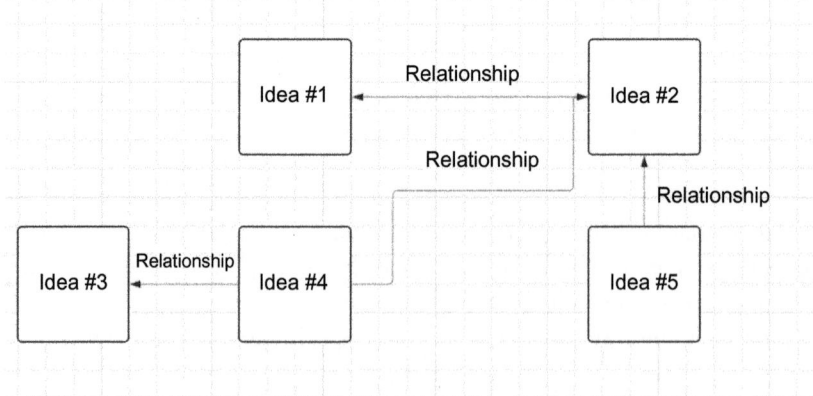

Figure 2.2 Concepts and relationships among them organised in a concept map

neatness. A concept map created with mind-mapping software may not be easily comprehensible. Concept mapping software is required to generate a concept map, just as mind mapping software is necessary for creating a mind map. Various software programmes, both online and offline, can be utilised to create mind maps and concept maps, either freely or through a registration charge. Additional information on tools and programmes for constructing mind maps can be found in this book.

Finally, how do we know when one has read all the necessary literature and when to stop?

Firstly, as emphasised by James Hayton, PhD,[13] the attempt to read everything ever written is discouraged due to its impossibility. The key lies in the selection of what to read and what to omit, avoiding an exhaustive approach. Although having a thorough understanding of the area is thought to be essential, it is usually thought to be sufficient to have a working knowledge of its basic concepts, procedures, and research kinds. With this broad knowledge in place, easy reference to the literature to retrieve specific works (or examples of work types for a literature review) is facilitated as needed.

The literature directly pertinent to your own work is the only portion necessitating in-depth scrutiny and detail. The critical aspect lies in offering a reflective analysis of the sources, explaining the motivation behind citing them, and avoiding the misconception of attempting to convey: 'Here is all there is to know about such a topic.' There is no obligation to incorporate every piece of information encountered in the literature review. The selected literary works should function as the most apt illustrations of the points intended to be conveyed. Depending on the relevance to your own work, the level of detail included in each segment of the literature review should be adjusted accordingly.

A valuable resource for contemplating these issues is the guidance provided by Erasmus Universiteit Rotterdam[14] on when to cease reading and commence writing (EUR Library—Erasmus University Rotterdam (n.d.)). To address the initial concern regarding how to gauge the sufficiency of literature examination, it is essential to recognise that there is no definitive definition of 'enough' to strive towards. Posing the question in this manner may not lead to a comprehensive literature review, even if an arbitrary target, such as 150 references, is set. Instead of fixating on a specific quantity, it is advisable to consider the depth required in each area and identify references that best exemplify your point or the type of research you are addressing. The key question should revolve around what you intend to communicate with your literature review. As a rule of thumb, I would say it all boils down to criteria for the inclusion or exclusion of sources, a well-crafted research question, and your supervisor's expertise.

---

[13] https://phd.academy.
[14] https://www.youtube.com/watch?v=ojRfybNJNuk.

Crafting a research question

Formulating a research question

I shall start this chapter by quoting Albert Einstein: 'If I had an hour to solve a problem, I'd spend 55 minutes thinking about the problem and five minutes thinking about solutions.'

Finding an interesting problem or gap in the literature is the first step in formulating a good research question. By good, I mean a research question that is relevant and interesting for the audience it is intended for, for example a community of practitioners, a group of researchers, academics, the general public, society, and so on.

All PhD students aspire to develop compelling theories with widespread appeal, sometimes even causing a paradigm shift. Formulating creative research questions that will lead to engaging and necessary studies is a crucial phase in developing any theory.

Gap-spotting is a common strategy for generating research topics, which requires a proper systematic literature review but leaves current literature primarily unchallenged. An interesting approach to formulating a proper research question is suggested by Alvesson and Sandberg (2013) in their book entitled *Constructing research questions: Doing interesting research*. Actual methods of developing research questions are frequently ignored or very briefly mentioned in traditional research methods courses. In their text, Alvesson and Sandberg, using examples from across the social sciences, develop a problematisation methodology for identifying and challenging the assumptions underlying existing theories, as well as for generating research questions that can lead to more intriguing and influential theories.

Crafting research questions can pose challenges, particularly for those who have not engaged in the process before.

A good research question includes the following elements, readapting from what Nikki Anderson[15] wrote about product research:

1. It is not focused on the solution but on learning something new about your audience or the community of researchers or practitioners you are interested in; instead of focusing only on results, a good research question seeks to understand something about the community of people revolving around it.
2. It's a challenge or notion that we don't entirely grasp. Make sure the research question fills a knowledge gap or is novel.
3. It relates to a concept or idea about which we need more details in order to proceed. As said above, we want to ensure that our research topic will

---

[15] https://dscout.com/people-nerds/guide-better-questions.

enable us to obtain the data necessary to advance an idea or make a better conclusion. Consider what your original contribution is in the field of interest.

This blog entry, entitled 'A step-by-step guide to writing better research questions', lists the steps for generating a valid research question, and many of those steps are opportunely transposed to research; I find it useful for a PhD student as well:

1. *Brainstorm.* What questions do you have, and where are the gaps in the theories and knowledge currently available? What discrepancies exist between your hypothesis and the data already available? What could you study to help you comprehend this gap better?
2. *Contribution.* What new information might the answers to these questions provide for your audience? Do these inquiries relate to the goals of the study? Will they provide the information you require to proceed to be facilitated by answering these questions?
3. *Focus and evaluation.* Which question is most crucial to obtaining the data you require? Does a quantitative or qualitative (or mixed-method) study need to be conducted to answer this question?
4. *Wording.* What is the appropriate language to describe a research question? Did you familiarise yourself with the appropriate semantics used in the reference community?
5. *Type of questions.* Variance questions are 'what' inquiries that are best answered by quantitative approaches since they deal with what happened and whether something happened due to another item (the relationship between variables or causality). Process inquiries are more suited to qualitative methodologies since they examine narratives and scenarios about a certain topic and explain 'why'.

To quote Bertrand Meyer here:

A researcher does not work 'in' an area but 'on' a question (Meyer, 2016). That is, when asked in which area you work, you have to be able to reply with a question you are tackling that can involve different areas. Beware of simplification when people tell you I work in this area or that; in fact, such a statement does mean nothing really specific in research. A real researcher does not just follow the flow, working 'in' a specific fashionable area or at the confluence of two fashionable areas. A real researcher attempts to solve open problems and so has to be able to describe a problem and a research question with a clear and concise statement.

Crucial guidance on how to write a proper research question comes from Van de Ven's book *The engaged scholarship* (2007) in chapter three on 'Formulating the research problem'.

One of the main lessons in identifying a relevant problem describes how many of us deal with complex issues beyond our limitations. When we move outside of ourselves and involve pertinent stakeholders in the learning process, we can better understand these issues than when we do so alone.

The approach suggests that, to formulate a proper research question, we need to involve a form of participatory inquiry. That is, a participatory research method where participants are brought in, and their various views are used to gain knowledge of a problem topic, a description of how academics characterise their interactions with their subject matter and communities, students, practitioners, and other academics.

To properly situate the problem, we will tackle a series of questions:

- Define the research's target audience and intended users. Who or what is the problem's foreground and background? A study on e-learning, for instance, can prioritise teachers over students or vice versa (which makes it a different problem).
- What is the problem domain's analysis level? Individuals, groups, practitioners, organisations, industries, or society may all be affected by the issue.
- What are the depth, breadth, and time dimensions of the issue? How long, how broadly, and how deeply should a problem be studied?
- Beware of the scope creep factor, mentioned by Van de Ven (2007), which means realising that the issue is much wider than anticipated and necessitates the involvement of stakeholders in its characterisation and diagnosis.

The journalistic pattern of 'who', 'what', 'when', 'where', 'why', and 'how' might come in handy for describing a problem and the corresponding research questions. An example, a story, or a personal experience can be utilised to illustrate the problem. Engaging with those affected by the issue is valuable; employing methods such as focus groups, ethnography, and interviews can extract knowledge. Participation in workshops dedicated to the topic of interest provides an opportunity to meet individuals dealing with similar issues. Additionally, a review of the literature can be conducted to ascertain whether a systematic literature review exists, describing the issues and future challenges related to the problem you aim to address.

As proposed by Van de Ven (2007), diagnosing a problem enables the identification of specific aspects that are of interest for study. Diagnosis is the process of putting observed phenomena into predetermined categories, choosing from a list of pre-listed answers. Without having the perfect response, you can try just to match the data you have collected about the problem.

You can aggregate observation and data into categories, just like categorising patients' data by symptoms. Use heuristic matching and triggers to aggregate data and link the problem to the solution. You can work out a description of the problem with different strategies, such as data-directed (from data to abstractions) or hypothesis-directed, that is, reversing a problem's direction (hypothesis) and assembling proof to back it (experimental method).

For instance, mixed methods (data and hypothesis-oriented) might lead to a focused search that produces new data that may lead to refocusing. You can also look for breakdowns, which occur when assumptions are broken, expectations are not satisfied, or something does not make sense. Indeed, to encourage discovery, you need a variety of different hypotheses.

Finally, when elaborating your research question from the problem you identified, write it by linking important concepts from the problem domain and phrase it in analytical terms. Link your problem description to the research question. Make sure your subject is a question and not a statement, and make sure it is of interest to your audience.

Good guidance on presenting your research question comes from a book entitled *Why some ideas survive and others die* (Heath and Heath, 2007). This book's main lesson is that any idea may be presented in a way that makes it memorable. Successful narratives, ad campaigns, and concepts that endure typically have recognised traits that can be encapsulated in the mnemonic SUCCESs:

- Simple—Simply go to the heart of any concept.
- Unexpected—pique interest by surprising the audience.
- Concrete—make an idea clear and understandable so that it can be retained in memory.
- Credible: able to convey plausibility.
- Emotional: Used to persuade people of the value of a concept.
- Story: using narrative to enable people to use an idea.

To summarise all the advice on how to write a proper research question, you can refer to a piece written by Shone McCombes on Scribbr in 2022.[16]

Precisely, what is desired to be learnt and shared through the effort is identified by a research question. Undoubtedly, a strong research question must guide the dissertation. The relevant characteristics of a research question have been listed by McCombes.

A research question must be:

- Focused on a single problem or issue

---

[16] https://www.scribbr.com/research-process/research-questions/.

- Researchable using primary and/or secondary sources[17]
- Feasible to answer within the timeframe and practical constraints
- Specific enough to answer thoroughly
- Complex enough to develop the answer over the space of a paper or thesis
- Relevant to your field of study and/or society more broadly

In summary, the exploration of the main research question is a common aspect of dissertations. The principal statement, formed in response, serves as the main argument in the dissertation. Throughout the course of your work, multiple research questions or issue statements are typically necessary to address the main research question. Breaking down the main research question into sub-questions is a common approach, often corresponding to specific sub-studies and, consequently, to chapters in your dissertation. However, these sub-questions must exhibit distinct relevance to one another and focus on a single study issue.

The phrasing of your question should align with the purpose of your research. Examples of how to construct queries for various contexts are provided in Table 2.1 following McCombes.[18]

Table 2.1 Research objective and corresponding formulation

Research objectives	Research question formulations
Describing and exploring	
	- What are the characteristics of X? - How has X changed over time? - What are the causes of X? - How has X dealt with Y?
Explaining and testing	
	- What is the relationship between X and Y? - What is the role of X in Y? - What is the impact of X on Y? - How does X influence Y?
Evaluating and acting	
	- What are the advantages and disadvantages of X? - How effective is X? - How can X be improved?

---

[17] https://www.scribbr.com/working-with-sources/primary-and-secondary-sources/.
[18] https://www.scribbr.com/research-process/research-questions/.

A final piece of advice is about how to break down a research question into sub-questions. Good sub-questions must be less difficult than the primary question, concentrated only on one kind of research and placed in a logical sequence.

Try to keep sub-questions to a maximum of three or four. If there is a feeling that more questions are needed, it may suggest that the core research topic is not sufficiently specific. In this situation, it is preferable to review the problem statement and attempt to make the key inquiry more concise.

Making an original contribution

Several ways exist in which new ideas can be generated for your PhD work after the completion of the literature review. Strategies that may be found helpful include:

- Gaps in the existing research should be sought as the literature is reviewed, with attention directed towards areas characterised by fewer studies or limitations in the existing research.
- Alternative perspectives should be considered, contemplating the existing research from different angles or viewpoints, such as examining a problem or issue from distinct scientific contexts.
- Connections between various studies or ideas in the literature should be identified, as these connections may indicate novel directions for research.
- Researchers, academics, or advisors should be engaged in discussions about research interests, seeking their input for potential new ideas or perspectives that may not have been previously considered.
- Taking a break from the research, if feeling stuck, can be beneficial. Returning to the work with fresh eyes after a hiatus may provide a new perspective and facilitate a renewed understanding of the subject matter.

The iterative process of generating new ideas requires time and effort, and it may be necessary to invest considerable resources before arriving at a research question or problem that elicits excitement for exploration in your dissertation.

Alan Dix (2004), in the 'Research techniques' notes, provides some good advice on how to make an original contribution. There are several heuristics that can be useful.

Starting the problem abstraction process can be undertaken by examining the problem and contemplating it as an illustration of a more extensive and widespread issue. Subsequently, an examination of connected issues associated with the problem can be conducted. The solutions pertinent to the larger issue can then be adapted to address the specific problem at hand. This method, involving abstraction and analogy, is commonly employed in various academic disciplines and proves valuable in real-world situations as well.

One of the widely acknowledged but challenging pieces of advice for coming up with original ideas is De Bono's lateral thinking (De Bono & Zibalist, 1970). The objective is to think differently from how one typically thinks. For instance, looking for various classification methods is a means of imposing different viewpoints. Another strategy for getting away from conventional solutions is to ask highly in-depth 'why' inquiries (Card, 2017).

As suggested by Dix (2004), continuously challenging yourself is another approach to fostering fresh thoughts. Imagine we only had one thought; we can try to figure out what's wrong with it. We can start by taking it apart and considering its advantages and disadvantages even if we believe it to be a fantastic idea. Such an approach will lead to rethinking our decision. To get around this, we can come up with absurd concepts. It is more comfortable criticising dumb ideas since we don't intend for them to be good ideas, which is a benefit. We can, however, also search for positives. Take a ridiculous notion and act as though it were something important, just like we did with other people's work. Why is it good or terrible? Its absurdity itself can encourage new connections and, in turn, new ideas (Dix et al., 2006).

Another good source for idea generation is exploring tools and techniques to foster innovation and creativity. The Hyperisland[19] website lists many different techniques and tools to consider.

Among the innovation toolkits, we might consider:

- *The how-now-wow matrix.* It is a tool used to evaluate and prioritise ideas or projects. It is often used in business or organisational contexts as a way to assess the potential impact and feasibility of different ideas or initiatives. The matrix consists of a grid with three categories: 'how', 'now', and 'wow'. Ideas or projects are evaluated based on how much they align with each of these categories. 'How' refers to the practicality and feasibility of an idea or project. This includes considerations such as the resources required, the time and effort needed to implement the idea, and any potential risks or challenges. 'Now' refers to the urgency of an idea or project. This includes whether the idea is timely and relevant and whether it should be pursued immediately or can wait for a later time. 'Wow' refers to the potential impact and value of an idea or project. This includes whether the idea has the potential to be transformative or make a significant difference in the organisation or its stakeholders. Ideas or projects that score highly in all three categories are considered to be the most promising and should be prioritised. Those ideas that score lower in one or more categories may still be worth pursuing but may require more careful consideration or additional resources to be successful.

---

[19] https://hyperisland.com/.

- *The 5-whys method.* The 5-whys method is a problem-solving technique used to identify the root cause of a problem by repeatedly asking the question, 'Why?' It is a simple and effective way to drill down to the underlying cause of a problem rather than just addressing the symptoms.

Here's how the 5-whys method works:

1. Clearly define the problem that needs to be addressed.
2. Identify the first cause of the problem by asking, 'Why did this happen?'
3. Identify the second cause of the problem by asking, 'Why did this cause happen?'
4. Repeat this process until the root cause of the problem is identified.

For example, let's say there is a problem with a machine in a factory that is causing production delays. Using the 5-whys method, the problem might be analysed as follows:

1. Why did the machine stop working? (Cause: A component failed)
2. Why did the component fail? (Cause: It was not properly maintained)
3. Why was it not properly maintained? (Cause: The maintenance schedule was not followed)
4. Why was the maintenance schedule not followed? (Cause: The maintenance team was not properly trained on how to follow the schedule)
5. Why was the maintenance team not properly trained? (Root cause: A lack of training in the organisation)

The 5-whys method is most effective when used as part of a team, as it encourages collaboration and diverse perspectives. It can be a useful tool for identifying and addressing problems in a variety of settings, including manufacturing, service industries, and even personal or professional development.

- *Idea and concept development.* This is the process of generating, refining, and evaluating ideas and concepts for new products, services, or initiatives. It is an important step in the innovation process, as it helps organisations identify and explore potential opportunities for growth and differentiation; this too might be employed successfully to generate your research question.

There are several stages involved in idea and concept development, including:

1. Idea generation: This involves generating a large number of ideas, either through brainstorming sessions, market research, or other methods like the

6-8-5 technique in which you generate 6–8 ideas in five minutes without taking too much time to reflect on the idea itself.

2. Idea refinement: This involves reviewing and refining the ideas generated in the first stage, eliminating those that are not feasible or viable, and selecting the most promising ideas to move forward.

3. Concept development: This involves creating a detailed plan or outline for the selected ideas, outlining the key features and benefits, target market, and potential risks and challenges.

4. Concept testing: This involves testing the concept with potential customers or stakeholders to gather feedback and validate the concept's viability.

Idea and concept development is an iterative process, and it is common for ideas and concepts to be refined or modified as they move through the different stages. It is essential to involve a diverse group of people in the process, as this can help generate a broader range of ideas and perspectives.

Among the creativity toolkits, various options can be considered:

1. *Start small.* Any creative endeavour has the potential to cause some worry. Who hasn't heard that the first step is always the hardest? Michael Avatar argues that we take inspiration from the Zen Buddhist notion of the 'beginner's mind', which holds that everything is only beginning in his book *Being creative*. A newbie attitude is crucial in today's society, according to futurist and author Kevin Kelly, because none of us have been experts for very long in a world where technology is constantly evolving. Therefore, it may be beneficial to realise that we begin every day and every moment. Possibility, openness, and curiosity are all characteristics that are advantageous for an investigation of creativity. During a creativity block or at the commencement of a workshop or project, this straightforward practice can be employed either individually or collaboratively with a group.

2. *Protobot.* Building something is one of the best ways to explore creativity. Simple, low-fidelity prototypes can help us turn simple (and occasionally sophisticated) ideas into tangible objects. By doing this, we unavoidably create room for more investigation, reflection, and iteration. We can discover the fun of hands-on experimentation with the aid of this tool. This is a fantastic exercise to do either alone or with others.

3. *Rapid research.* A quick activity that adds outside perspectives and viewpoints to exploratory, discursive, and creative workshops. Use this activity when coming up with fresh ideas, coming up with a new product or service, or coming up with a strategy or plan that will involve other people. Participants contact someone in the research community of interest and ask them questions about the problem they are tackling. This generates valuable

feedback from a variety of 'outside' perspectives quickly. Participants are frequently taken aback by how easy it was to get this feedback and how important it is to the process.

Theoretical framework versus conceptual framework

PhD students' research is often built on a theoretical and/or conceptual framework (Rockinson-Szapkiw & Spaulding, 2014). These frameworks are explored in this chapter because they clarify the study's goal and serve as a template for developing research questions, which in turn assist students in selecting the best approach. In other words, the study's conceptual framework—the author's synthesis of the theoretical, seminal, and empirical literature on a topic—must be in line with the research problem, the objective of the investigation, and the research questions. There is also a discussion of additional factors to take into account while choosing a methodology.

An idea or concept that offers a broad, abstract explanation of a phenomenon or group of phenomena is known as a theoretical framework. When formulating research hypotheses or interpreting research results, a theoretical framework is frequently based on accepted theories and concepts in the subject matter. On the other hand, a conceptual framework is a more focused set of concepts and ideas that provide a framework for comprehending and examining a particular fact or series of phenomena. A conceptual framework often serves to organise and direct the collection and analysis of data in a research study. It frequently builds upon or extends a theoretical framework.

In the social and behavioural sciences, among others, theoretical and conceptual frameworks are crucial tools because they offer a way to arrange and make sense of complicated phenomena and to create hypotheses or justifications for research findings. They are utilised at various stages of the research process and for various reasons. Conceptual frameworks offer a more focused structure for comprehending and interpreting a particular occurrence than theoretical frameworks, which offer a more general, abstract explanation of a phenomenon.

In 'right-sizing your research' method, Rockinson-Szapkiw and Spaulding (2014) offer extensive guidance, including a case study on how to develop both a theoretical and conceptual framework. The conceptual framework should demonstrate two things: (a) how the doctorate student's research relates to existing theory and research and (b) how the student's research offers a contribution to the area (its intellectual goals). In essence, the conceptual framework for the study is provided by the literature review. The conceptual framework's objective is not to synthesise prior work in the field. Rather, it is to establish your planned study's foundation in the significant prior work and to clearly convey to the reader your theoretical approach to the phenomena you propose to study. As mentioned by Rockinson-Szapkiw and Spaulding (2014), Ravitch and Riggan (2012) illustrate a conceptual framework as 'an argument about why the topic one wishes to study matters, and why the means proposed to study it are appropriate and rigorous. By argument, we mean that a conceptual framework is a series of sequenced, logical propositions, the purpose of which is to convince the reader of the study's importance and rigour'. Understanding how the theoretical and conceptual frameworks help narrows the focus and provides justification for the research problem, the proposed research questions and the study's objectives, and the doctoral student's identification and justification of the types of knowledge required to answer those questions are crucial.

A theoretical framework usually consists of one or sometimes two well-known ideas, whereas a conceptual framework incorporates both theoretical and empirical literature that is utilised to guide the research. 'Any empirical or quasi-empirical theory of social and/or psychological processes, at a number of levels (e.g. grand, midrange, and explanatory), that can be applied to understanding a reality' is how Anfara and Mertz (2006) describe a theoretical framework. The theoretical framework provides an overarching explanation for how and why one would expect one variable to explain or predict another variable, informing the research questions and hypotheses, and in quantitative studies, where theory is used deductively and placed towards the beginning of the research plan. As a result, the theoretical framework plays a crucial role because the study is meant to evaluate the theory. In qualitative studies, the researcher advances a theory at the outset of the study, gathers data to test it, and then analyses the findings to determine if the theory was confirmed or disconfirmed (i.e. rejects or fails to reject the null hypotheses).

Choosing the appropriate research method

Selecting a research method

After a proper research question has been crafted and written down into concise and clear statements, it is necessary to explore which research methods available in the literature are the most appropriate to tackle the question. The main research methods in science and engineering include:

1. *Empirical methods*: Conducting experiments and observations to collect data and test hypotheses.
2. *Theoretical methods*: Deductive reasoning and mathematical models are used to provide proofs and explanations.
3. *Semi-empirical methods*: Combining empirical data with theoretical models such as the method of parameter variation.

4. *Quantitative methods*: Utilising surveys, experiments, simulations, and statistical analysis to measure and analyse data.

5. *Qualitative methods*: These methods employ narrative research, ethnography, phenomenology, grounded theory, and case studies to explore and describe phenomena.

6. *Mixed methods*: Integrating both qualitative and quantitative approaches to provide a more comprehensive understanding.

7. *Participatory methods*: Involving stakeholders and end users in the research process.

The choice of research method depends on the purpose of the study, in other words, the type of scientific research,[20] whether exploratory, descriptive, explanatory, or evaluative in nature. PhD students in science and engineering must select appropriate, reliable, and feasible methods that align with their research questions and objectives. As a rule of thumb, validity and reliability are used to assess the quality of research methods. They show the accuracy with which a procedure, test, or method measures something. Validity concerns a measure's precision, whereas reliability concerns a measure's consistency, as explained in Fiona Middleton's article on Scribbr.[21] The data type needed to address the research topic will also guide the choice of the appropriate research methods. Quantitative methods are utilised to quantify something or test a hypothesis, while qualitative methods aim to uncover the meanings people attribute to social phenomena, making them ideal for exploring concepts, experiences, and perceptions. Examples of qualitative methods include ethnographic fieldwork, in-depth interviews, focus groups, and document analysis, all of which provide valuable insights into diverse perspectives and cultural contexts. Secondary data is utilised if a large amount of easily accessible data needs to be studied. Primary data collection is preferred if customised data tailored to specific needs and control over its production are desired. Experimental techniques are used to establish cause-and-effect relationships between variables. On the other hand, descriptive approaches are employed to gain an understanding of the qualities of a study subject.

At this stage, it is crucial to assemble a range of techniques given the type of information that is intended to be used or collected. Then, one or a mix of research methods must be selected and elucidated to explain why they are the most suitable for achieving the research goal. It is also essential to consider the preferences of different communities. Different communities may have varying inclinations towards specific research methods based on their cultural norms, values, and research traditions. Understanding and respecting these preferences can enhance

---

[20] https://courses.lumenlearning.com/suny-hccc-research-methods/chapter/chapter-1-science-and-scientific-research/.

[21] https://www.scribbr.com/methodology/reliability-vs-validity/.

the relevance and acceptance of the research within the community. Therefore, PhD students should engage with the research advisory board, supervisor, and the community to ensure that the chosen research methods align with the community's expectations and needs, fostering a collaborative and culturally sensitive research approach.

A general guideline for choosing between qualitative and quantitative data is to use quantitative research to verify or test a hypothesis (a theory or hypothesis) and qualitative research to learn more about a subject (concepts, thoughts, experiences). A qualitative, quantitative, or mixed techniques approach can be chosen for the majority of study issues. The selection of the appropriate approach is contingent upon several factors, including the research question posed, the nature of the research design (whether experimental, correlational, or descriptive), and practical considerations such as temporal constraints, financial resources, data accessibility, and participant availability.

Quantitative and qualitative research allows you to address many types of research issues by using various data collection and analysis techniques, as shown in Table 2.2.

It's crucial to remember that both quantitative and qualitative research have advantages and disadvantages and that it's frequently beneficial to combine the two methods in a single study. Various techniques are available for obtaining both

Table 2.2 Quantitative versus qualitative research

Characteristic	Quantitative research	Qualitative research
Purpose	To test hypotheses and make predictions	To understand and interpret social phenomena
Research questions	Specific, focused	Broad, open-ended
Data collection	Large sample sizes, standardised instruments (e.g. surveys, experiments)	Small sample sizes, non-standardised instruments (e.g. interviews, observations)
Data analysis	Statistical techniques	Interpretive techniques (e.g. thematic analysis, content analysis)
Generalisability	A representative sample allows for generalisation to a larger population	Limited generalisability due to the small sample size and specific context
Reliability	High reliability due to standardised methods and a large sample size	Lower reliability due to non-standardised methods and a small sample size
Validity	External validity (ability to generalise) is high, but internal validity (ability to establish cause and effect) may be limited.	Internal validity is high, but external validity may be limited.

quantitative and qualitative data. A data collection technique should be chosen to address the research topic. Quantitative and qualitative data collection techniques are numerous. Data can be presented as numbers, utilising methods such as rating scales or counting frequencies, or as words in surveys, observational studies, or case studies, with open-ended questions or descriptions of observations.

However, neither qualitative nor quantitative data can prove or demonstrate anything on their own; instead, they need to be examined to reveal their significance in relation to the research questions.

Empirical methods

Empirical methods are fundamental to the scientific process in science and engineering. They provide a systematic approach to collecting data and testing hypotheses. These methods involve conducting experiments and observations to gather evidence that can be used to support or refute scientific theories. By relying on empirical evidence, researchers can make informed conclusions and contribute to the advancement of knowledge in various fields.

Empirical methods play a crucial role in scientific inquiry by providing a means to validate or falsify hypotheses. Through experiments and observations, researchers can gather data that is directly relevant to the questions they seek to answer. This empirical evidence serves as the foundation for scientific theories and helps to ensure the reliability and validity of research findings.

Experiments are a key component of empirical research. They allow researchers to manipulate variables and observe their effects on outcomes. Experimental research methods allow researchers to establish causal relationships between variables by controlling for confounding factors and employing randomisation techniques. Confounding factors are variables that may influence the relationship between the independent and dependent variables, potentially leading to spurious associations. Examples of confounding factors include age, gender, socio-economic status, or pre-existing conditions. By controlling for these factors, either through statistical techniques or experimental design, researchers can isolate the true effect of the independent variable on the dependent variable. Randomisation techniques involve randomly assigning participants to different experimental conditions or treatment groups. This helps ensure that any observed differences between the groups are due to the independent variable and not due to systematic differences in the characteristics of the participants. Common randomisation techniques include simple random sampling, block randomisation, and stratified randomisation.

By controlling for confounding factors and using randomisation, experiments enable researchers to make robust causal inferences about the relationships between variables. This allows them to confidently determine whether changes in the independent variable directly cause changes in the dependent variable rather than being influenced by other extraneous factors. This rigorous approach helps to minimise bias and ensure the accuracy of the results obtained.

Observational studies involve the systematic observation and recording of phenomena in their natural settings. They allow researchers to gather data on behaviours, patterns, and relationships, providing valuable insights into complex phenomena.

Hypothesis testing is one of the primary objectives of empirical research, as it aims to validate or refute hypotheses derived from theoretical frameworks. By formulating clear and testable hypotheses, researchers can design experiments or observational studies to gather data that either supports or refutes their predictions. This process of hypothesis testing is essential for advancing scientific knowledge and refining existing theories, so the next section is specifically on hypothesis testing.

Empirical methods require careful planning and execution of data collection procedures. Researchers must use reliable and valid measures to ensure the quality of the data obtained. Once data is collected, it must be analysed using appropriate statistical techniques to draw meaningful conclusions. Data analysis allows researchers to identify patterns, relationships, and trends that can inform their interpretations.

While empirical methods are valuable for generating evidence-based knowledge, they also have limitations. Certain phenomena may be difficult to study using traditional experimental or observational approaches. Additionally, biases and errors in data collection and analysis can impact the validity of research findings. Researchers must be aware of these limitations and take steps to mitigate them in their studies.

Empirical methods are essential for conducting rigorous and systematic research in science and engineering. PhD students can collect data to test hypotheses and validate theories by conducting experiments and observations. The use of empirical evidence helps to ensure the reliability and validity of research findings, contributing to the advancement of knowledge and understanding in science and engineering. PhD students must carefully design and execute empirical studies to generate high-quality data that can inform scientific enquiry and drive innovation.

Hypothesis testing

There are many statistical techniques that can be used to analyse quantitative data collected in the form of variables by observational studies in empirical settings. Some common techniques include:

- Descriptive statistics: Descriptive statistics involve summarising the key characteristics of a dataset, such as the mean, median, mode, standard deviation, and range.

- Inferential statistics: Inferential statistics involve making inferences about a population based on a sample. Common inferential statistics techniques include hypothesis testing, regression analysis, and analysis of variance (ANOVA).
- Multivariate analysis: Multivariate analysis involves analysing the relationship between multiple variables simultaneously. Common multivariate techniques include factor analysis, cluster analysis, and discriminant analysis.

Hypothesis testing is probably the most notable quantitative research method and deserves further consideration. For an introduction, refer to 'Introduction to hypothesis testing', chapter 8[22] of published by Sage (Privitera, 2016). See also some basic guidance given in an online tutorial by Duquesne University Library.[23]

Hypothesis testing is a statistical technique for evaluating the validity of a hypothesis by collecting and analysing data. The history of hypothesis testing can be traced back to the early twentieth century when researchers such as Ronald Fisher and Jerzy Neyman developed statistical methods (Lehmann, 2011).

Fisher is credited with developing the null hypothesis, a statement that there is no relationship or difference between variables. He also introduced the significance level, which is the probability of rejecting the null hypothesis when it is actually true. Fisher's work laid the foundation for modern statistical hypothesis testing.

Neyman and Egon Pearson, on the other hand, developed the idea of the alternative hypothesis, which is a statement of the relationship or difference between variables that are being tested (Neyman & Pearson, 1933). They also introduced the concept of the power of a test, which is the probability of correctly rejecting the null hypothesis when it is false.

Today, hypothesis testing is a widely used statistical technique in many fields, including psychology, economics, and the social sciences. It is used to test hypotheses about the relationships between variables and to decide the validity of a research question or theory.

Using statistics, hypothesis testing is a formal process for examining our theories about the world. Scientists most frequently employ it to examine particular hypotheses that result from theories (Wheelan, 2013).

To test a hypothesis, follow these five steps:

---

[22] Find this chapter online: https://us.sagepub.com/sites/default/files/upm-binaries/40007_Chapter8.pdf.
[23] https://guides.library.duq.edu/c.php?g=844215&p=6035683.

1. Create a null hypothesis (H_0) and an alternate hypothesis for your research (H_a or H_1).
2. Sample data so that it will help you test the theory.
3. Run the relevant statistical tests.
4. Make a decision regarding whether to accept or reject the null hypothesis.
5. The findings are presented in the results and comments section.

The procedure employed to test a hypothesis typically adheres to a variation of these phases, notwithstanding potential alterations in specific methodologies.

(1) Hypothesis. To facilitate quantitative testing, it is imperative to reformulate the initial research hypothesis, which is the hypothesis under investigation, into both a null hypothesis (H0) and an alternative hypothesis (H1). The alternative hypothesis generally posits a relationship between the variables, whereas the null hypothesis posits no correlation between the variables of interest. Suppose we are trying to determine whether a new teaching method is effective at improving student test scores. We start by selecting a random sample of students and dividing them into two groups: a treatment group that will receive the new teaching method and a control group that will receive the traditional teaching method. Our null hypothesis (H_0) is that the new teaching method has no effect on test scores, and our alternative hypothesis (H_1) is that the new teaching method does have an effect on test scores.

(2) Sampling data. Conducting sampling and collecting data in a manner conducive to evaluating the hypothesis is imperative for ensuring the legitimacy of a statistical test. Failure to obtain representative data may preclude drawing accurate statistical conclusions about the population of interest. The student sample should be distributed evenly across genders and encompass diverse socio-economic backgrounds. Additionally, it should incorporate any pertinent control variables that could potentially impact average performance, thereby facilitating the examination of variations in test scores. Moreover, consideration should be given to the scope of the study. Utilising data from the census, which provides comprehensive information across various geographic locations and includes factors such as income levels and parental education, could serve as a valuable resource in this context.

(3) Run a statistical test. Several statistical tests are available, but they are all focused on contrasting between-group variance and within-group variance (how dispersed the data are within a category and how different the categories are from one another). Between-group variance, sometimes referred to as intergroup or among-group variance, measures how much the means of the several study groups vary. It evaluates the degree of variation in results or reactions among different experimental conditions or participant groups by assessing the variability between the group averages. Increased variation within groups indicates a higher degree of dissimilarity in the responses or performance of the groupings. Conversely,

within-group variance, which is also called intragroup variance or residual variance, quantifies the degree of variability among individual data points in every condition or group. Indicating the degree to which individual data points depart from the group mean, it evaluates the spread or dispersion of scores within the same group. Greater variability or heterogeneity among individual observations within each group is indicated by a higher within-group variance. While within-group variance evaluates variability within each group, between-group variance concentrates on variances between the group means. Understanding the general pattern of data dispersion and interpreting the findings of statistical analyses, such as ANOVA or regression models, depend on these variance components.

The statistical test will indicate a low p-value (Wheelan, 2013) when the between-group variance is sufficiently large to minimise overlap among groups. Consequently, the likelihood of these group differences arising by chance is remote. Conversely, a high p-value will be evident in the statistical test when there is notable within-group variance and minimal between-group variance. This suggests that any observed differences among groups are likely attributable to chance. The choice of variables and the level of measurement of the collected data will dictate the appropriate statistical test to be employed.

(4) *Accept or reject your null hypothesis.* The decision to reject or fail to reject the null hypothesis hinges upon the outcomes of the statistical test. Typically, this determination is guided by the p-value generated by the test. Given that a p-value below 0.05 implies a less than 5% probability of observing the obtained results if the null hypothesis were valid, the conventional significance level for rejecting the null hypothesis is often set at 0.05. However, researchers may opt for a more stringent significance criterion, such as 0.01 (1%), to mitigate the risk of erroneously rejecting the null hypothesis (type I error).

(5) *Present the findings.* In the results and discussion sections of the thesis, the outcomes of hypothesis testing are typically analysed and interpreted. The results section should encompass a succinct presentation of the data and a summary of the statistical test findings, including key metrics such as the estimated difference between group means and the associated p-value. Subsequently, in the discussion section, the implications of these findings are elaborated upon, providing insights into whether the initial hypothesis was corroborated by the results or not.

Finally, we would not normally state whether we accept or reject the alternative hypothesis. This is so because hypothesis testing doesn't aim to support or refute any claims. Its sole purpose is to determine if a pattern we observe may have happened accidentally or spuriously. We can argue that our test supports our hypothesis if our research leads us to reject the null hypothesis (i.e. we conclude that it is implausible that the pattern occurred by chance). However, if the pattern fails our decision criteria, indicating that it might have developed by chance, we can only declare that the test is incompatible with our hypothesis.

Theoretical methods

Theoretical methods are a cornerstone of scientific inquiry, providing a framework for developing and testing hypotheses, explaining phenomena, and predicting outcomes. These methods rely on deductive reasoning and mathematical models to give proofs and explanations that underpin our understanding of the natural world. By employing theoretical approaches, researchers can explore complex relationships, uncover underlying principles, and advance knowledge in various fields of science and engineering.

One example is catastrophe theory, a branch of mathematics used to model and analyse discontinuous behaviour in complex systems.

As described (Laymon, 1989), catastrophe theory is 'an elegant example of discontinuous behaviour' that can be used in engineering applications. Researchers in this area start with established mathematical principles and theories, such as the work of Zeeman (1972) and Poston and Woodcock (1973), as mentioned in Laymon (1989).

Using deductive reasoning, hypotheses are derived from how certain engineering systems or components might exhibit sudden, dramatic changes in behaviour or performance. Theoretical models and mathematical representations are developed to predict and explain these discontinuous phenomena. For instance, a researcher might hypothesise that a structural element in a bridge design will experience a sudden failure mode under certain loading conditions. She can then construct a theoretical model using catastrophe theory to predict when and how this failure might occur and derive the mathematical proofs to support her hypothesis. By testing these theoretically derived predictions against empirical data from experiments or observations, the researcher can provide explanations for the underlying mechanisms driving the discontinuous behaviour. This allows researchers to refine the theoretical framework and improve the predictive power of the model.

Theoretical methods, including deductive reasoning and mathematical modelling, are widely used in engineering research to develop hypotheses, explain complex phenomena, and forecast outcomes. They complement the empirical and semi-empirical approaches also employed in the field.

Theoretical methods are essential for building a solid foundation of knowledge in scientific disciplines. By using deductive reasoning and mathematical models, researchers can formulate hypotheses, derive predictions, and test the validity of theories. Theoretical methods help to organise and structure information, identify patterns, and make sense of complex phenomena. Without theoretical frameworks, scientific inquiry would lack coherence and direction, hindering our ability to explain and predict natural phenomena.

Theoretical methods are based on deductive reasoning, which is a logical process in which specific conclusions are drawn from general principles or premises.

In theoretical research, deductive reasoning is used to derive predictions from theoretical frameworks and testable hypotheses. By starting with established principles and applying logical reasoning, researchers can make inferences about the outcomes of experiments or observations. Deductive reasoning allows scientists to build upon existing knowledge and develop new insights into the workings of the natural world.

For example, in quantum computing, deductive reasoning plays a crucial role in developing algorithms and analysing quantum systems. Quantum computing leverages the principles of quantum mechanics to perform computations that are exponentially faster than classical computers for certain tasks (Hahanov et al., 2018).

Imagine a quantum computing researcher aiming to develop a quantum algorithm for factorising large numbers efficiently, a problem that is notoriously hard for classical computers but can be solved efficiently using Shor's algorithm on a quantum computer.

The researchers start by formulating a hypothesis based on the principles of quantum mechanics and the mathematical framework of quantum computing. They hypothesise that by leveraging quantum superposition and entanglement, they can design an algorithm that can factorise large numbers exponentially faster than classical algorithms.

Using deductive reasoning, the researchers explain how quantum superposition allows quantum bits (qubits) to exist in multiple states simultaneously, enabling parallel computation. They also explain how quantum entanglement links qubits together, allowing for correlations that classical systems cannot achieve.

The researchers predict that by applying Shor's algorithm (Shor et al., 1994), a quantum computer with a sufficient number of qubits can efficiently factorise large numbers into their prime factors. The outcomes of running the algorithm on a quantum computer are anticipated through deductive reasoning and mathematical modelling, predicting the speed-up compared to classical factorisation methods.

Another technique is to use mathematical models, which are powerful tools used in theoretical research to represent and analyse complex systems and phenomena. These models use mathematical equations, algorithms, and simulations to describe relationships, make predictions, and test hypotheses. By quantifying variables and parameters, researchers can explore the behaviour of systems under different conditions and scenarios. Mathematical models provide a quantitative framework for understanding phenomena that may be difficult to study through empirical methods alone.

For instance, in the field of climate science, mathematical models are powerful tools used to represent and analyse complex systems and phenomena. Climate models are essential for understanding the Earth's climate system, predicting future climate trends, and assessing the impact of various factors on the environment.

Climate scientists use mathematical models to develop hypotheses about how the Earth's climate system functions (Czocher et al., 2014). They hypothesise how factors like greenhouse gas emissions, solar radiation, and ocean currents interact to influence global temperatures and weather patterns. Mathematical models in climate science explain a wide range of phenomena, such as the greenhouse effect, El Niño events, and sea-level rise. These models help scientists understand the underlying mechanisms driving climate change and variability.

Climate models predict future climate scenarios based on different emission scenarios and policy interventions. By simulating the Earth's climate system using mathematical equations and algorithms, researchers can forecast temperature changes, precipitation patterns, and extreme weather events.

In this example, mathematical models are used in climate science to represent the complex interactions within the Earth's climate system. These models incorporate equations that describe the dynamics of the atmosphere, oceans, land surface, and ice sheets. By running simulations and analysing the outputs of these models, scientists can make predictions about how the climate will evolve under different scenarios.

Theoretical methods aim to provide proofs and explanations for observed phenomena by establishing logical connections between concepts and principles. Through deductive reasoning, researchers can demonstrate the validity of their hypotheses and theories by showing how they logically follow established premises. Mathematical models help to formalise these relationships and provide a quantitative basis for testing predictions and hypotheses. Researchers can offer rigorous proofs and explanations for natural phenomena by combining deductive reasoning with mathematical modelling.

These frameworks consist of concepts, principles, and relationships that help to explain and predict phenomena. By constructing theoretical models, researchers can propose hypotheses, derive predictions, and test the validity of their theories. Theoretical frameworks provide a roadmap for exploring complex phenomena and guiding research efforts in a systematic manner.

One key aspect of theoretical research is testing the predictions derived from theoretical models. By comparing the outcomes of experiments or observations with the predictions of these models, researchers can evaluate the validity of their theories. This process of hypothesis testing helps refine theoretical frameworks, identify areas for further investigation, and validate the underlying assumptions of the models. Testing theoretical predictions is essential for building confidence in the reliability and accuracy of scientific theories.

Theoretical methods have a wide range of applications across various scientific disciplines, including physics, chemistry, biology, engineering, and the social sciences. In physics, theoretical methods are used to develop models of the universe, predict the behaviour of particles, and explain fundamental forces. In chemistry, theoretical approaches help to understand molecular structures, chemical reactions, and thermodynamic properties. In biology, theoretical methods are employed to study evolutionary processes, ecological systems, and genetic interactions. In engineering, theoretical models are used to design and optimise complex systems, predict performance outcomes, and solve practical problems. Theoretical methods also play a crucial role in social sciences by providing frameworks for understanding human behaviour, societal trends, and cultural dynamics.

While theoretical methods offer valuable insights into the natural world, they also face challenges and limitations. Developing accurate and predictive theoretical models can be complex and time-consuming, requiring sophisticated mathematical techniques and computational tools. Theoretical assumptions may oversimplify real-world phenomena, leading to inaccuracies in predictions and explanations. Additionally, theoretical models may be limited by the availability of data, the complexity of systems being studied, and the uncertainty inherent in scientific inquiry. PhD students must be aware of these challenges and limitations when using theoretical methods in their work.

As scientific knowledge continues to expand and evolve, theoretical methods will play a crucial role in shaping our understanding of the world around us. Advances in computational modelling, data analytics, and artificial intelligence are opening up new possibilities for theoretical research. By integrating theoretical methods with empirical approaches, researchers can gain deeper insights into complex phenomena, uncover hidden patterns, and make novel discoveries. Theoretical methods will continue to drive innovation, inspire new research directions, and contribute to the growth of scientific knowledge in the years to come.

Semi-empirical methods

Semi-empirical methods represent a specific approach to scientific research, blending empirical data with theoretical models to enhance the accuracy and efficiency of analyses. One prominent example of semi-empirical methods is the method of parameter variation, which involves adjusting specific parameters within a theoretical model to align with empirical data.

Semi-empirical methods serve as a bridge between empirical observations and theoretical frameworks, offering a middle ground that combines the strengths of both approaches. By integrating empirical data with theoretical models, researchers can refine their understanding of complex phenomena and improve the predictive power of their analyses. The method of parameter variation is a key technique within semi-empirical methods, allowing researchers to adjust model parameters to better fit empirical data.

The method of parameter variation involves systematically adjusting specific parameters within a theoretical model to optimise its alignment with empirical data (Stewart et al. 1989). This iterative process allows researchers to fine-tune the model's predictions and enhance accuracy. By varying parameters such as constants, coefficients, or initial conditions, scientists can calibrate the model to better reflect real-world observations.

The method of parameter variation is widely used in scientific and engineering research across various disciplines.

For example, in computer science, semi-empirical methods are instrumental in optimising machine learning algorithms and models. By blending empirical data with theoretical frameworks, researchers can fine-tune algorithm parameters to enhance model accuracy, efficiency, and predictive power. This approach allows for the development of more robust and effective machine-learning solutions that can adapt to changing data patterns and improve overall performance.

One notable application of semi-empirical methods in computer science is machine learning (Hu et al., 2023), where these methods optimise algorithms and models based on empirical observations. In machine learning, researchers often hypothesise that certain algorithm parameters can be adjusted to improve the performance of a model. Researchers can develop hypotheses about how specific parameters impact model accuracy and efficiency by combining empirical data from training sets with theoretical insights into algorithm behaviour.

Semi-empirical methods in machine learning help explain how adjustments to algorithm parameters influence model outcomes. By analysing empirical data and theoretical models, researchers can gain insights into the underlying mechanisms that drive algorithm performance and make informed decisions about parameter optimisation.

Using semi-empirical methods, researchers can predict the outcomes of adjusting algorithm parameters on model performance. By iteratively refining parameters based on empirical data and theoretical insights, they can make accurate predictions about how these adjustments will impact the efficiency and accuracy of machine-learning models.

Semi-empirical methods, including the method of parameter variation, offer several advantages in scientific research. By combining empirical data with theoretical models, researchers can enhance the accuracy and reliability of their analyses. This approach also allows for the incorporation of real-world observations into theoretical frameworks, leading to more robust predictions and explanations.

However, semi-empirical methods also present challenges. The process of parameter variation can be computationally intensive and time-consuming, requiring careful calibration and validation. Additionally, the success of semi-empirical methods hinges on the quality and relevance of the empirical data used and the appropriateness of the theoretical model being adjusted.

Quantitative methods

Quantitative research methods lie at the heart of science and engineering empirical inquiry. They involve systematically collecting and analysing numerical data to understand phenomena, test hypotheses, and derive insights that can guide technological innovations.

Based on the techniques used to acquire it, data can be divided into four basic categories: observational, experimental, simulational, and generated, as mentioned by the Dewit Wallace Library research guide.[24] The nature of the research data gathered may influence how the data is managed. For example, additional backup methods are essential for data that is difficult or impossible to replace, such as recording events at precise times and locations, to mitigate the risk of data loss. Conversely, adherence to recommended practices is necessary to prevent data corruption when integrating data points from various sources.

Through the observation of a behaviour or action, observational data are gathered. Data is gathered by techniques including casual observation, in-depth questionnaires, or using a device or sensor to track and record data of persons or objects on the move and supply a timely ordered sequence of location data for further processing. An example of casual observation is a researcher sitting in a café and observing the interactions between customers and baristas. They take note of body language, tone of voice, and other non-verbal cues to understand the dynamics of customer service in the establishment. In-depth questionnaires, such as surveys administered to university students, can be used to understand students' attitudes towards online learning. The questionnaire might include detailed questions about their experiences with online courses, preferences for learning formats, and challenges they face with remote education. Data can be collected from a tracking device, such as a fitness tracker worn by individuals during their daily activities, which records steps taken, heart rate, and sleep patterns. The device automatically collects and logs this data, which users can access through a companion app to track their fitness goals and monitor their health metrics over time. Because observational data is recorded in real time, losing them would make re-creating it challenging or even impossible, which makes a data backup strategy crucial.

When a variable is changed, experimental data is gathered actively by the researcher to cause and measure change. Experimental data is frequently projectable to a broader population and typically enables the researcher to identify a causal relationship. Collecting these kinds of data is repeatable, but doing so is frequently expensive. For example, in testing the impact of a new teaching method, an educational researcher experiments to evaluate the effectiveness of

---

[24] https://libguides.macalester.edu/c.php?g=527786/&p=3608643.

a new teaching method in improving student learning outcomes in mathematics. Two groups of students from similar demographics are selected: one group receives instruction using the traditional teaching method (control group), while the other group receives instruction using the new teaching method (experimental group). The researcher actively collects data by administering pre-tests and post-tests to assess students' mathematical proficiency before and after the instructional intervention. Additionally, the researcher gathers data on students' engagement, motivation, and attitudes towards mathematics. The experimental data obtained from this study allows the researcher to project the findings to a broader population of students and to establish a causal relationship between the teaching method and its impact on learning outcomes. However, conducting such experiments in educational settings can be costly due to the need for specialised training, resources, and data collection and analysis time.

By employing computer test models to simulate the behaviour of a real-world process or system over time, simulation data is produced. For instance, it can be used to forecast earthquake activity, chemical processes, economic models, or weather conditions. This approach is used to try to predict what might occur given particular circumstances. The data produced by the simulation is frequently just as significant as, if not more so than, the test model that was utilised. This data provides a greater knowledge of the behaviour of the system under various conditions in addition to validating the simulation model's accuracy and dependability. Furthermore, the data produced by simulations frequently enables researchers to investigate and evaluate circumstances that would be difficult or unethical to repeat in actual studies. As a result, it is essential for policy formation, decision-making processes, and the advancement of scientific knowledge in various fields, including hazard mitigation and earthquake preparedness, chemical process optimisation, economic policy, and weather forecasting.

Generated data has undergone some sort of modification, such as an arithmetic formula or aggregation, from previously existent data points, frequently from various data sources. For instance, population density data is created by integrating area and population data from two similar cities, or financial ratios, such as return on investment (ROI) or debt-to-equity ratio, are generated from financial statements by applying specific formulas to existing data points, such as revenue, expenses, assets, and liabilities. While it is typically possible to recreate this type of missing or deleted data, doing so could be very time-consuming and costly. In this section, we will delve into various quantitative research methods available based on the nature of data collection.

Surveys: surveys when involving close-ended questions in the form of Likert scale options, for example 'On a scale of 1 to 5, please rate your level of agreement with the following statement: "I feel satisfied with the quality of customer service provided by our company". Likert scale options: 1—Strongly disagree,

2—Disagree, 3—Neutral, 4—Agree, 5—Strongly agree. Surveys can be used to ask a standardised set of questions to a large sample of people. Surveys can be administered in person, by phone, or online.

Experiments: experiments involve manipulating one or more variables and measuring the effect on a dependent variable. Experiments allow researchers to establish cause-and-effect relationships. For example, if a researcher wants to investigate whether the amount of time students spend studying affects their performance on a maths test, the researcher will recruit a group of participants and randomly assign them to two conditions: a control group and an experimental group. The control group will be instructed to study for one hour, while the experimental group will be instructed to study for three hours. After the designated study period, all participants will take the same maths test. The researcher records the test scores for each participant. The researcher compares the average test scores between the control group (one-hour study time) and the experimental group (three-hour study time). If the experimental group performs significantly better on the test compared to the control group, it suggests that increased study time has a positive effect on test performance. Based on the results, the researcher can conclude that there is a cause-and-effect relationship between study time and test performance. Specifically, increasing study time leads to improved performance on maths tests.

Observational studies: observational studies involve observing and measuring behaviour or characteristics in a natural setting without manipulating variables. For example, if a researcher wants to understand employee communication patterns in an office environment.

The researcher spends time in the office, discreetly observing employees' communication behaviours during various tasks, meetings, and interactions. The researcher will refrain from intervening or influencing the natural flow of communication. Using field notes or a structured observation checklist, the researcher records details such as the mode of communication (e.g. face to face, email, phone), the frequency of interactions, the duration of conversations, and the participants involved. After multiple observation sessions, the researcher analyses the data to identify communication patterns within the office. Trends are examined, such as preferred communication channels for different types of tasks, the role of hierarchy in communication dynamics, or differences in communication styles between departments. The researcher can draw conclusions about workplace communication patterns based on the observations. For example, face-to-face communication might be discovered to be more prevalent among employees in creative departments than in administrative departments. Observational studies provide valuable insights without manipulating any variables.

Secondary data analysis: secondary data analysis involves analysing data collected by someone else, such as government agencies or other research studies.

For instance, a researcher aims to investigate energy consumption patterns in manufacturing plants to identify opportunities for energy efficiency improvements. The researcher obtains secondary data from government energy agencies and previous research studies. This data includes information on energy usage (such as electricity consumption) and production output collected from various manufacturing plants over several years. The researcher analyses the collected secondary data to examine energy consumption patterns across different manufacturing plants and industries. Statistical techniques may be used to identify trends, seasonal variations, and factors influencing energy usage, such as production levels, operational schedules, and equipment types. Based on the secondary data analysis, the researcher may find that certain manufacturing processes or industries exhibit higher energy consumption rates compared to others. For instance, peak usage periods or inefficiencies in energy utilisation might be examined. These insights can help inform strategies for optimising machine usage, implementing energy-saving measures, and reducing environmental impacts in manufacturing operations.

Longitudinal studies: longitudinal studies involve collecting data from the same participants over an extended period of time, allowing researchers to study changes or trends over time.

For instance, a researcher aims to investigate students' academic achievement from primary school through to college to understand the factors influencing academic performance over time. The researcher recruits a cohort of students from several primary schools and follows them longitudinally over the course of their education. Data collection begins when the students are in year one and continues annually until they graduate from college. Each year, the researcher collects data on various factors related to academic achievement, including standardised test scores, grades, attendance records, socio-economic status, family background, extracurricular activities, and educational interventions. The same set of participants is assessed consistently over the entire study period. The researcher analyses the longitudinal data to examine changes or trends in academic achievement over time. Statistical methods such as growth curve modelling or hierarchical linear modelling might be used to assess individual trajectories of academic progress and identify predictors of academic success. Based on the longitudinal study, academic growth patterns or decline among the participants over the years might be spotted. They may identify factors such as parental involvement, socio-economic status, and early literacy skills that significantly influence academic achievement trajectories.

Panel studies: panel studies involve collecting data from the same group of participants at multiple points in time. For example, a researcher aims to investigate the health behaviours and outcomes of adolescents over a period of several years to understand the long-term effects of lifestyle choices on health and well-being. The researcher recruits a panel of adolescents from diverse backgrounds and follows them longitudinally from adolescence into young adulthood.

Participants in the panel study are assessed at multiple time points using surveys, interviews, and medical examinations to collect data on their health behaviours, such as diet, physical activity, substance use, and mental health, as well as health outcomes, such as obesity, cardiovascular health, and mental well-being. The same group of participants is assessed consistently over the study period. The researcher analyses the panel data to examine changes or trends in health behaviours and outcomes over time. Statistical methods such as growth curves, latent class analysis, and trajectory analysis might be employed to identify patterns of behaviour change and their associations with health outcomes. Based on the panel study, the researcher may observe trajectories of health behaviour adoption and changes in health outcomes among the participants as they transition from adolescence to young adulthood. Factors such as peer influence, socio-economic status, and access to healthcare that shape health behaviours and outcomes over time can be analysed.

Meta-analysis: meta-analysis involves combining the results of multiple studies to reach a larger conclusion. A group of researchers aims to evaluate the overall effectiveness of Mobile Health (mHealth) interventions in supporting diabetes management by synthesising multiple studies conducted in this area. The researchers conduct a systematic literature review to identify relevant studies investigating the impact of mHealth interventions on diabetes-related outcomes such as glycemic control, medication adherence, and quality of life. Studies that utilise mobile apps, wearable devices, or other digital tools to deliver interventions to improve diabetes self-management can be conducted. The researchers extract key data from each selected study, including study design, intervention components, participant characteristics, duration of follow-up, and outcome measures. They also record effect sizes or relevant statistics reported in the original studies. Using statistical methods, such as random-effects meta-analysis, the researchers combine the results of the selected studies to calculate summary effect sizes and confidence intervals. They explore heterogeneity across studies and conduct subgroup analyses to examine the influence of factors such as intervention type, duration, and participant demographics on intervention effectiveness. The researchers might find that mHealth interventions are associated with significant improvements in diabetes management outcomes compared to standard care or alternative interventions. They identify specific components of mHealth interventions, such as real-time feedback, goal-setting features, and educational content, that contribute to their effectiveness.

This list represents only the main quantitative data collection methods, but other methods in the literature might be more suitable depending on the research questions a PhD student wants to tackle.

Qualitative methods

Qualitative research methods encompass a diverse range of approaches that delve into the depth and complexity of human experiences, behaviours, and phenomena. These methods, including interviews, focus groups, ethnography, and case studies, among others, offer a rich and nuanced understanding of the social, cultural, and psychological aspects of the subjects under study. By focusing on the subjective meanings and interpretations of individuals, qualitative methods provide valuable insights that quantitative approaches may not capture. In this section of the book, we will delve into various qualitative research methods available based on the nature of data collection.

Interviews: interviews involve asking open-ended questions and allowing the participant to provide in-depth responses. Interviews can be conducted in person, by phone, or online. For example, a question such as 'Can you describe a situation where you had to learn a new software program or tool to accomplish a task, and how did you go about mastering it?' A response can demonstrate the candidate's ability to adapt to new software environments, proactively seek knowledge and resources, and persistently overcome challenges to achieve proficiency. It will highlight their self-directed learning skills and willingness to collaborate with others to enhance their capabilities.

Focus groups: focus groups involve bringing together a small group of people to discuss a specific topic in depth. For example, a university is considering implementing a new online learning platform to supplement traditional classroom instruction. They want to gather feedback from both students and faculty members to ensure that the platform meets their needs and expectations. Different phases are required to gather and analyse data using focus groups:

Planning and recruitment: the university identifies students from various academic programmes and faculty members from different departments who have experience with online learning platforms. Invitations to participate in the focus group are sent out via email and posted on university bulletin boards, with information about the purpose of the session and the expected time commitment.

Set-up: the focus group session is scheduled at a convenient time for participants, either during a break between classes or in the evening. A comfortable and quiet meeting room on campus is selected as the venue, equipped with audiovisual aids if needed.

Moderator: a moderator, possibly a faculty member or a staff member from the university's instructional design team, introduces themselves and explains the purpose of the focus group. Participants are informed that their feedback will be crucial in shaping the university's decision regarding the adoption of the new online learning platform.

Discussion: The same moderator begins by asking open-ended questions to prompt discussion among participants, such as 'What are your experiences with online learning platforms?' 'What features do you find most valuable in an online learning environment?', and so on.

Participants share their perspectives, insights, and suggestions based on their experiences. The moderator encourages participants to provide specific examples and elaborate on their opinions.

Feedback collection: throughout the discussion, the moderator takes notes or records the session to capture key points and insights. Participants may also be asked to complete brief surveys or evaluations to provide additional feedback on specific aspects of the online learning platform.

Wrap-up: towards the end of the session, the moderator summarises the main themes and takeaways from the discussion. Participants are thanked for their valuable contributions and may receive a small token of appreciation, such as a gift card or refreshments. The university commits to analysing the feedback collected from the focus group and using it to inform their decision-making process regarding the implementation of the new online learning platform.

By conducting a focus group with students and faculty members, the university gains valuable insights into their preferences, concerns, and expectations regarding online learning, allowing them to make informed decisions that support effective teaching and learning practices.

Observations: observations involve watching and recording behaviour or events as they occur in a natural setting. Observations can be structured (following a predetermined set of rules) or unstructured (allowing the researcher to follow the conversation or events as they unfold).

For instance, a research team is studying energy usage patterns in households to identify opportunities for energy conservation. They plan to use Internet of Things (IoT) devices, such as smart meters and sensors, to collect real-time data on electricity consumption and household activities.

Set-up and data collection: the research team installs IoT devices, including smart meters, motion sensors, and temperature sensors, in participating households. Smart meters are installed to measure electricity consumption at regular intervals, providing granular data on energy usage patterns. Motion sensors are placed in key areas of the house, such as the living room, kitchen, and bedrooms, to detect human activity and movement. Temperature sensors are deployed to monitor indoor temperature variations throughout the day.

Structured observations: researchers may conduct structured observations by defining specific parameters and rules for data collection. For example, they may set predetermined time intervals (e.g. every hour) to record energy usage data from smart meters and sensor readings from motion and temperature sensors; for this particular case, researchers might follow a structured protocol for observing and recording household activities, such as the use of appliances, lighting, and heating or cooling systems.

Unstructured observations: in addition to structured observations, researchers also engage in unstructured observations to capture spontaneous events and behaviours as they occur. For instance, IoT technology can be used to passively monitor activities and events without interfering with the natural flow of household routines. In such cases, researchers may observe changes in energy consumption patterns in response to external factors such as weather conditions, time of day, or occupants' behaviour.

Data analysis and insights: a research team analyses the collected data to identify trends, correlations, and anomalies in energy usage patterns. They examine how household activities and behaviours influence electricity consumption and indoor environmental conditions. Insights from the observations help researchers understand the factors driving energy consumption and inform the development of personalised recommendations for energy-saving strategies.

By combining structured and unstructured observations with IoT technology, the research team gains valuable insights into energy usage behaviours and patterns in households, enabling targeted interventions for energy conservation and sustainability.

Case studies: case studies involve in-depth analysis of a single individual, group, or situation.

For example, a case study examines the implementation of a remote patient monitoring (RPM) system within a healthcare organisation to improve the management of chronic diseases, focusing on patients with hypertension. The objective of this case study is to evaluate the effectiveness of implementing an RPM system in facilitating remote monitoring, early detection of hypertension-related complications, and personalised care delivery.

Selection of participants: a cohort of patients diagnosed with hypertension is identified from the healthcare organisation's patient database. Patients are invited to participate in the RPM programme based on predefined eligibility criteria, including diagnosis of hypertension, access to a smartphone or internet-enabled device, and willingness to participate.

Implementation: a user-friendly RPM platform is deployed, allowing patients to measure and transmit their blood pressure readings, medication adherence, and lifestyle data remotely. Patients receive personalised instructions on how to use the RPM devices and are provided with ongoing technical support as needed.

Data collection and monitoring: patients are instructed to measure their blood pressure according to a predefined schedule using the RPM devices. Blood pressure readings are automatically transmitted to the healthcare provider's electronic health record (EHR) system for real-time monitoring and review. Healthcare providers conduct regular virtual consultations with patients to discuss their blood pressure trends, medication adherence, lifestyle modifications, and any potential concerns.

Evaluation and outcomes: qualitative feedback from patients and healthcare providers is obtained through surveys, interviews, and focus groups. Key outcomes evaluated include improvements in blood pressure control, patient satisfaction, adherence to treatment plans, and healthcare provider efficiency.

The case study demonstrates the effectiveness of implementing an RPM system for chronic disease management, particularly in the context of hypertension. By leveraging digital health technologies, healthcare organisations can enhance patient engagement, improve clinical outcomes, and optimise resource utilisation in the delivery of care.

Ethnographic studies: ethnographic studies involve immersing oneself in a culture or community and observing and participating in the daily life of the group. For example, an ethnographic study of 'User interaction with a social media platform interface'. This study aims to gain insights into how users interact with the interface of a popular social media platform in their daily lives, focusing on their behaviours, preferences, and challenges.

Participant selection: participants are recruited from diverse demographics, including age, gender, and geographic location, to ensure a representative sample. Informed consent is obtained from each participant, clarifying the purpose of the study and the nature of their involvement.

Immersion in user environment: researchers immerse themselves in the daily lives of participants, observing and participating in their interactions with the social media platform across different contexts and settings. Observation sessions take place in participants' homes, workplaces, and other relevant environments where they typically access the social media platform.

Participatory observation: researchers actively engage with participants as they navigate the interface of the social media platform, asking questions, seeking clarification, and encouraging open dialogue. Participants are encouraged to narrate their actions, thoughts, and experiences while interacting with the platform, providing rich insights into their decision-making processes and motivations.

Contextual inquiry: researchers pay close attention to the contextual factors that influence user interaction with the interface such as environmental distractions, time constraints, and social dynamics. They observe how users adapt their behaviours and navigation strategies based on situational cues and constraints encountered in their daily lives.

Data collection: observations are documented through field notes, audio recordings, and visual media, capturing both verbal and non-verbal cues. Screenshots or screen recordings may be captured to supplement observational data and provide a contextual understanding of interface usage patterns.

Data analysis: qualitative data analysis techniques, such as thematic coding and narrative analysis, are employed to identify recurring themes and patterns in user interaction with the interface. Emergent themes may include usability challenges, feature preferences, navigation strategies, and emotional responses to interface design elements.

By adopting an immersive and participatory approach, researchers can uncover nuanced behaviours, preferences, and challenges that inform interface design and usability considerations.

It's important to note that these are just a few examples of qualitative data collection methods, and there are many others as well.

Mixed methods

There are many research methods that can be used in a plethora of fields, from humanities to sociology to information systems, when your research involves humans and technology (Lazar et al., 2017). Some common methods include:

1. Survey research involves collecting data from a sample of individuals using self-report measures or structured interviews. Surveys can be administered in person, by mail, or online.
2. Experimental research involves manipulating an independent variable and measuring its effect on a dependent variable. This allows researchers to establish cause-and-effect relationships between variables.
3. Case study research involves an in-depth investigation of a single case or a small number of cases. It is often used to explore complex issues or phenomena in depth.
4. Action research involves conducting research in a real-world setting and working with practitioners to develop solutions to problems or to improve processes.
5. Grounded theory research involves collecting and analysing data to generate a theory that explains the phenomena under study.
6. Ethnographic research involves studying a culture or group of people through observations, interviews, and other forms of qualitative data collection.
7. Design science research is a research approach focused on the design, creation, and evaluation of artefacts, systems, and processes intended to solve problems or meet needs in a specific domain.

These are the main among the numerous research techniques that can be applied in the information systems sector. The research question being addressed and the resources available to the researcher will determine the most acceptable strategy.

(1) Survey research. This is a technique for gathering information from a sample of people using questionnaires or other self-reporting tools (Arleck et al., 1995).

Survey research, which can be carried out online, over the phone, or in person, is frequently used to examine attitudes, beliefs, and behaviours. The ability to quickly and effectively gather data from a large number of people is one of the key benefits of survey research. Surveys can be given to a sample of people that is typical of the entire population, enabling researchers to draw conclusions about the population from the sample's responses. Survey research does have certain limits, though. Self-report measures, for instance, may be prone to biases like the social desirability bias, where people may not disclose their real opinions or behaviours in order to present themselves in a more favourable light. Additionally, survey research depends on people's willingness and capacity to participate, which may have an impact on the sample's representativeness. In general, survey research is a useful method for gathering information and gaining an understanding of attitudes, beliefs, and actions, but it is crucial to carefully evaluate its limitations and the possibility of bias in the results.

(2) Experimental research. One or more independent variables are manipulated in experimental research to see how they affect a dependent variable. The aim of experimental research is to establish cause-and-effect correlations between the variables being researched (Campbell & Stanley, 2015). In experimental research, control and experimental groups are frequently used. The control group, which is not subjected to the modification of the independent variable, acts as a reference point against which the experimental group can be contrasted. On the other hand, the experimental group is subjected to the manipulation of independent variables. Researchers can ascertain the impact of the manipulation on the dependent variable by comparing the results of the control and experimental groups. In the natural and social sciences, experimental research is frequently used to test hypotheses and theories about how various variables may be related. Considering that it enables researchers to establish cause-and-effect linkages, it is regarded as a very rigorous and dependable approach to study. However, experimenter bias and the challenge of recreating the experimental conditions in real-world contexts are two limitations of experimental research.

(3) Case study research. This is a technique for gathering and analysing data in depth, frequently regarding a single person, group, or society (Simons, 2009). In order to conduct case studies, a variety of data concerning the case must be gathered, including both quantitative (such as statistics or measurements) and qualitative data (such as observations, interviews, and documents). Studies of rare or unusual occurrences can benefit greatly from the use of case studies, which are frequently utilised to examine complicated issues or phenomena in their natural settings. They can offer comprehensive and in-depth accounts of the experiences and viewpoints of the parties concerned and aid academics in comprehending the situation and setting. There are several different ways to perform case study research, including interviews, observations, and document analysis. It is crucial for researchers to give considerable thought to the techniques they employ

and to guarantee the validity and reliability of the data they gather. Overall, case study research is a useful method for delving deeply into complicated problems and occurrences, but it's crucial to carefully evaluate its drawbacks, including the possibility of bias and the difficulty of extrapolating the results to a larger population.

(4) Action research. This is a type of inquiry that involves people in the study process with the goal of bringing about good social change. Action research comprises a collaborative and iterative process of problem identification, data collection, analysis, and action planning and is frequently employed in educational, community development, and organisational settings (Stinger, 1996). The fact that the study subjects actively participate in the research process is one of the main characteristics of action research. This may entail co-designing the study, gathering and evaluating data, and developing solutions to the investigated issues. The fact that action research is iterative and frequently involves numerous cycles of data collection and analysis, which produces new insights and knowledge, is another important characteristic of action research. Several factors need to be taken into account when doing action research. Setting up specific goals and objectives for the research is crucial, as it includes all pertinent parties in the investigation. Additionally, it is crucial to thoroughly organise and carry out the research, including data collection and analysis, and to make sure the results are communicated to all pertinent parties.

(5) Grounded theory research. In order to create a theory that adequately describes the phenomenon being examined, researchers use a method called grounded theory, which entails the methodical collecting and analysis of evidence (Birks & Mills, 2015). In the social and behavioural sciences, grounded theory—which comprises an iterative process of data gathering, coding, and analysis—is frequently employed to examine complex social processes and phenomena. Grounded theory is also used in computer science in the exploration of user experiences and behaviours in interactive systems. By collecting and analysing qualitative data, such as user interviews, observations, and system interactions, researchers can develop theories grounded in the empirical evidence gathered from real-world contexts. This approach allows for the identification of patterns, themes, and underlying mechanisms that shape user interactions with technology. Additionally, grounded theory is employed, for instance, in software engineering research to investigate the processes, practices and challenges associated with software development and maintenance. Through iterative data collection and analysis, researchers can uncover emergent themes and theoretical constructs that elucidate the complexities of software development practices, team dynamics, and project management strategies. Grounded theory is inductive, which means that the theory develops from the evidence rather than being imposed on it. This is one of its fundamental characteristics. This implies that the researcher starts with a broad research question or problem, gathers and analyses evidence, and then

develops a theory to account for the phenomenon under study. The researcher compares the data to existing theories and concepts to spot patterns and links, another important aspect of grounded theory. When conducting grounded theory research, there are a number of things to keep in mind. The research must be carefully planned and carried out, including the data collection and processing, and the careful interpretation and reporting of the results. Additionally, it is crucial to make sure that the theory developed is based on the patterns and relationships found in the data and that the research is founded in the data.

(6) Ethnographic research. A type of data collection and analysis known as ethnographic research entails the systematic observation and interpretation of the culture, behaviours, and beliefs of a specific group or community. The social and behavioural sciences frequently use ethnographic research to examine how individuals interact with one another and their surroundings. For example, an application of ethnographic research in engineering is in the study of user needs, preferences, and behaviours in the design of technological artefacts. By observing and engaging with users in their natural environments, engineers can gain a deep understanding of how individuals interact with technology, identify usability challenges, and uncover unmet needs that traditional methods may overlook. This user-centred approach to design helps engineers create more intuitive, effective, and user-friendly products and systems. Ethnographic research also plays a crucial role in the study of technology adoption and implementation within organisational settings. By conducting ethnographic studies within engineering firms, research laboratories, or manufacturing facilities, engineers can investigate how technologies are integrated into work processes, assess the impact of technological interventions on workflow and productivity, and identify barriers and facilitators to successful technology deployment. This ethnographic perspective enables engineers to develop more contextually relevant and sustainable engineering solutions tailored to the specific needs and challenges of organisational stakeholders. This type of research requires a thorough awareness of the culture being studied (Atkinson, 2007). The researcher participates in and observes the culture being researched to get a thorough grasp of the group or society's customs, principles, and values. This immersion is a major component of anthropological research. Another important aspect of ethnographic research is its interpretive nature, which involves the researcher trying to comprehend the importance and meaning of the data from the participant's point of view. When undertaking ethnographic research, there are a number of things to keep in mind. The research must be carefully planned and carried out, including data collection and processing, as well as the careful interpretation and reporting of results. Additionally, it is crucial to make sure that the research is respectful, ethical, and attentive to the culture and beliefs of the community or group being investigated.

(7) Design science research. The goal of design science research is to develop and assess artefacts, products, systems, or procedures that address issues and enhance

the quality of life for people. Design science research involves creating and testing ideas, models, and frameworks that can be used to design and enhance real-world systems (Hevner & Chatterjee, 2010). It is frequently utilised in domains including information systems, engineering, and management. The fact that design science research is iterative and involves numerous cycles of design, development, and evaluation in order to enhance the artefact, system, or process under study is one of its main characteristics. Design science research is action-oriented with the aim of developing useful solutions that can be applied in actual settings, which is another important characteristic of the field. When undertaking design science research, there are a number of things to bear in mind. It is crucial to characterise the issue or opportunity being addressed precisely and state the research's aims and objectives. Additionally, it's critical to determine and assess the right success metrics and to make sure that the results are shared in a way that others can use and improve upon the research.

Participatory methods

Participatory methods in science and engineering involve actively engaging stakeholders and end users in the research process. This approach consists of giving importance to considering the needs, perspectives, and experiences of those who will ultimately benefit from or be impacted by the research outcomes. By incorporating participatory methods, researchers can ensure that their work is relevant, effective, and sustainable. Participatory methods offer several benefits, such as involving stakeholders and end users, which can improve relevance and impact. This ensures that their work addresses real-world problems and meets the needs of those who will use the research outcomes. This increases the likelihood of the research having a positive impact and being adopted in practice.

Participatory methods can reduce the time and resources required for research by identifying the most critical problems and solutions earlier in the process. This is particularly important in fields where resources are limited or where rapid innovation is needed.

Participatory methods foster collaboration between researchers, stakeholders, and end users. This leads to a more diverse range of perspectives and expertise, which can result in more innovative and effective solutions.

Participatory methods provide researchers with a deeper understanding of the context and needs of the stakeholders and end users. This is critical in fields where the research is applied to real-world problems, as it ensures that the solutions are tailored to the specific needs and constraints of the users. These methods can increase the legitimacy and acceptance of research outcomes among stakeholders and end users. This is particularly important in fields where the research is applied to sensitive or controversial issues, as it can help build trust and credibility.

Participatory methods are being applied in various fields of science and engineering. For instance, *Healthcare*, where they are being used to develop new medical devices and treatments that are tailored to the needs of patients and healthcare providers. Researchers have involved patients in the design of prosthetic limbs to ensure that they meet the needs and preferences of those who will use them.

In *Environmental Sustainability*, participatory methods are used to develop sustainable solutions to environmental problems. For example, researchers have involved local communities in the design of renewable energy systems to ensure that they are culturally and environmentally sustainable.

In *Transportation*, participatory methods are used to develop more efficient and sustainable transportation systems. For instance, researchers have involved commuters in the design of public transportation systems to ensure that they meet the needs and preferences of those who will use them.

Participatory methods are being used in *Agriculture* to develop more sustainable and productive agricultural systems. For example, researchers have involved farmers in the design of new agricultural technologies to ensure that they are practical and effective.

While participatory methods offer many benefits, they also present several challenges, including being time-consuming and resource-intensive, particularly if they involve extensive stakeholder engagement and co-design activities.

Participatory methods can be challenging when significant power imbalances exist between researchers and stakeholders or end users. This can lead to unequal participation and influence in the research process. These methods often require the development of new or adapted research methods, which can be complex and require significant expertise. Participatory methods can also be vulnerable to co-optation, where stakeholders or end users are drawn into the research process primarily to legitimise the research rather than to contribute meaningfully.

Summarising, by involving stakeholders and end users in the research process, researchers can develop solutions that meet the needs of those who will use them, increase the efficiency and effectiveness of the research, and enhance collaboration and knowledge sharing. While participatory methods present several challenges, the benefits they offer make them an important tool for researchers seeking to make a positive impact in their fields. As the complexity and interdisciplinarity of research continue to grow, the importance of participatory methods will only increase.

Summary

This chapter offers a comprehensive exploration of essential research methods, guiding readers through the critical stages of the research process. It begins by emphasising the importance of conducting a systematic literature review, providing practical advice on developing inclusion and exclusion criteria for sources, creating effective search strategies, and critically reading academic materials. The chapter offers valuable insights into structuring a literature review, ensuring researchers build a solid foundation for their work. The chapter delves into the art of formulating proper research questions, outlining a strategic approach to identify compelling aspects of a problem and developing focused inquiries. Practical steps are provided to help researchers refine their questions, ensuring they are well crafted and conducive to meaningful research. The chapter then transitions into a discussion of various research methods available in the literature, assisting readers in selecting the most appropriate approaches for their chosen topics. This includes an informative comparison of quantitative and qualitative methods, enabling researchers to make informed decisions about their methodological framework. By addressing these fundamental aspects of research methodology, the chapter equips readers with the necessary tools and knowledge to embark on their research journey with confidence, from the initial stages of literature exploration to defining research questions and selecting suitable methods.

3

From dissertation to publication

Dissertation

Planning to write a dissertation

A PhD student's commencement of the dissertation writing process marks a critical turning point in their academic career. The dissertation provides an opportunity to exhibit intellectual curiosity and make a novel research contribution to the field of study of choice. Making sure it works requires careful preparation and careful consideration of many factors, which makes the project difficult. In particular, I would suggest that you reflect on these items:

- It is never too early to start planning and outlining for a dissertation. Ideally, one should start working on a dissertation proposal as soon as possible, and once it has been accepted, one should begin writing the dissertation.
- It's crucial to divide a PhD dissertation into smaller, more achievable tasks because it might be a huge undertaking. Consider establishing a schedule or project plan that divides the writing process into more manageable, smaller goals (see the 60:30:10 rule,[1] for example).
- It is advisable to create an outline for the dissertation before commencing the writing process. This practice aids in structuring ideas and ensuring the coherence and logical flow of the dissertation.
- Setting reasonable goals is paramount when evaluating the volume of work achievable per day and devising a schedule for dissertation completion. It is important to factor in any additional commitments, such as research or teaching responsibilities, into your planning process (see SMART goals[2]).
- To establish a consistent writing routine, consider creating a structured schedule. Allocate a set number of hours each day dedicated to dissertation work or aim to produce a specific word count on a weekly basis. Maintaining such a schedule can help sustain motivation and enhance focus throughout the writing process.
- During the dissertation writing process, it is advisable to solicit feedback from various sources. Advisors, peers, or other members of the academic

---

[1] https://www.larksuite.com/en_us/topics/productivity-glossary/60-30-10-rule.
[2] https://www.atlassian.com/blog/productivity/how-to-write-smart-goals.

The Doctorate Blueprint. Alessio Malizia, Oxford University Press. © Alessio Malizia (2025).
DOI: 10.1093/9780198927167.003.0004

community may provide constructive criticism. Incorporating such feedback can enhance the quality of the dissertation and improve the clarity of ideas presented.

- To prevent burnout, it's crucial to rest while writing. Schedule time for socialising, exercising, and engaging in other activities that will keep you energised and motivated.

How to write a dissertation

Producing a dissertation is a major undertaking that calls for thorough planning, close attention to detail, and a lot of effort. A candidate's ability to do independent research and advance knowledge in their field of study is to be shown through their dissertations. Along with a discussion of the research methodology and the relevance of the findings, a well-written dissertation should also offer a clear and succinct summary of the findings.

Selecting an engaging topic that aligns with the field of study serves as the first step in dissertation writing. Identifying a topic that strikes a balance between novelty and existing literature can pose a challenge. Following topic selection, conducting an exhaustive literature analysis is essential. This process aids in pinpointing key open-ended questions and knowledge voids that the thesis aims to address.

The subsequent phase in the dissertation writing process involves crafting a precise and concise research topic or hypothesis. This should constitute a question that can be empirically answered, one that is specific and focused in scope. Additionally, it is essential to develop a comprehensive research plan delineating the procedures and methods for data collection and analysis. This plan should encompass details regarding the sample, data-gathering instruments and procedures, and the statistical analysis techniques to be employed.

Subsequent to data collection and analysis, findings should be presented in a clear and concise manner. This entails providing a comprehensive explanation of the procedures and methods utilised, along with a summary of the results and their pertinence to the research topic. Recommendations for further research should be included, alongside an acknowledgement of any study limitations. In the dissertation conclusion, major conclusions should be summarised, and their broader implications should be explored. Furthermore, the contributions of the research to the field of study should be discussed.

Good guidance, including how to start writing a dissertation, can be found in 'How theses get written: Some cool tips!'[3] published by Dr Steve Easterbrook from the University of Toronto.

---

[3] https://www.doc.ic.ac.uk/research/phd/thesiswriting.pdf.

The text offers instructions on how to write a PhD thesis. The planning, structure, writing, and editing processes are all covered in detail. Additionally, it offers guidance on literature reviews, findings presentation, and citations. The slides emphasise the value of concise writing and offer pointers on how to do so. Additionally, it offers information on how to prevent frequent pitfalls and examples of both excellent and bad practices. Overall, the manual provides an extensive tool for PhD candidates who want to produce a strong thesis. One day, Desmond Tutu stated, 'There is only one way to eat an elephant: a bite at a time.'[4] He was saying that anything in life that seemed difficult, overpowering, or even impossible could be done gradually by taking things little by little. More specifically, the guidance offered by Dr Easterbrook in slide five addresses exactly how to start writing today, which includes a piece of advice about the following:

- Decide your title
- Write your title page
- Start a binder
- (Look at some thesis/dissertation in your area)
- Plan your argument
- You can change and refine this later, but at least you have a starting point (the first elephant's bite)

On slide seven, the presentation provides a template, replicated here in Table 3.1, to assist in planning the argument, which is considered very useful. Finally, on slide 10, advice is provided on how to utilise the template: converting the argument into a chapter outline, initiating a binder with a division for each chapter, and remaining open to revisions while working on the chapters.

Draft the dissertation structure

The following template, Table 3.2, is a good starting point for drafting the dissertation structure whenever readiness permits. It is recommended that you discuss it with the supervisor first. Some numbers referred to as metadata might be set by the institution or department, for example the minimum number of pages or the word limit.

The length and structure of a PhD dissertation might vary depending on the programme's specific requirements and the student's research. However, here's a ballpark figure: for example, a PhD dissertation in information systems and computing is typically 150 to 300 pages long. However, this can vary depending on the nature of the research, the amount of data collected, and the institution's formatting rules. A PhD dissertation typically ranges from 60,000 to 100,000

---

[4] https://www.psychologytoday.com/intl/blog/mindfully-present-fully-alive/201804/the-only-way-to-eat-an-elephant.

Table 3.1 Plan your argument

One sentence for each	Example
Introduction (area of study)	
The problem (that I tackle)	
What the literature says about this problem	
How I tackle this problem	
How I implement my solution	
The result	

Table 3.2 Typical research paper structure

Suggested page count	Chapter	Intended contents	Notes
5–10 pp	Introduction	Brief description	Specific notes
20 pp	Literature review	Brief description	Specific notes
15 pp	Research methodology	Brief description	Specific notes
15 pp	Prototype	Brief description	Specific notes
15–20 pp	Study1	Brief description	Specific notes
15–20 pp	Study2	Brief description	Specific notes
15–20 pp	Study3	Brief description	Specific notes
10 pp	Conclusions	Brief description	Specific notes
20 pp	References		

words. However, the word count may vary based on the study undertaken and the institution's formatting rules. Typically, a dissertation has 100 to 200 references. The amount of references, however, can vary depending on the nature of the research. It is vital to note that when it comes to references, quality is more important than quantity, so the number of references should not be the primary focus. Depending on the length and complexity of the thesis, as well as the level of detail required for the figures and the number of references cited, a typical PhD thesis may have anywhere from 20 to 50 or more pages of references and figures.

Draft dissertation—structure—<Your Name>

Metadata (example): assume references and figures are around 60 pages. The number of pages per text is 150 pp, @ 300 words per page = 45,000 words (total number of words). The approximate number of references is 160.

It's essential to adapt this table according to the specific structure and requirements of your dissertation, ensuring it aligns with your research objectives and methodology. Discussing this proposed structure with your supervisor can provide valuable insights and guidance for refining and finalising your dissertation outline.

Dissertation title

Following Dr Easterbrook's advice, the initial step is to begin drafting a title for your dissertation. The topic of academic titles becoming more difficult to understand and pronounce is covered in the article 'Why academics choose useless titles'[5] (Dunleavy, 2014).

That is a problem that, in my experience as a supervisor and examiner, plagues dissertations, too.

According to the author, this trend is motivated by a desire to stand out and set oneself apart in the competitive academic market. However, this emphasis on originality frequently comes at the sacrifice of accessibility and clarity, making it challenging for readers to comprehend what a paper or a dissertation is really about. The usage of unusual titles, according to the author, is widespread in the academic world and is not exclusive to any one discipline. However, in the social sciences, where writers frequently use titles that are dense with jargon and academic terminology, this trend is especially pervasive. This, according to the author, is a squandered opportunity because clear titles can draw in more readers and boost the effect of the study. According to the paper, researchers and PhD students should be more careful when selecting titles and try to find a balance between originality and intelligibility. The author advises that names should be succinct, direct, and clearly explain the research's major point. This will help to boost the visibility and impact of research while also making it simpler for readers to understand what a publication is about. The article concludes by pointing out that the usage of unusual titles is influenced by academic culture as much as personal preference. According to the author, academic institutions should try to foster a culture that prioritises clear and understandable communication and should offer tools and training to aid scholars in honing these abilities. By doing this, we can make sure that academic research is not just thorough and useful but also open to a larger audience and simple to grasp.

The article concludes by offering four suggestions for enhancing a title. The first stage is to come up with at least 10 potential titles and then compare them, making sure that the chosen words are not too ambiguous or conventional, and taking into account how readers would perceive them. The second phase is the use of consistent, linking words in the title, abstract, and subheadings. The third stage is to think about utilising a full narrative title that encapsulates the essence or main contribution of the argument. This can work well for articles and chapters.

---

[5] https://blogs.lse.ac.uk/impactofsocialsciences/2014/02/05/academics-choose-useless-titles/.

The fourth and last stage isto include at least a few narrative clues in the title, giving readers helpful cues or indicators about the thesis or line of reasoning.

Finally, the article offers examples of titles that are catchy and simple to quote, including 'New public management is dead—long live digital era governance', which sums up the paper's whole thesis and raises relevant questions. According to the paper, narrative signals in the title can still be useful even if a complete narrative header is too obvious.

Writing an introduction to your dissertation

The introduction should be prepared later after a thorough comprehension of the dissertation has been obtained. Fortunately, the article 'On the dissertation: How to write the introduction'[6] (Cassuto, 2018), taken from the website the Chronicle of Higher Education,[7] which in itself is a useful resource, comes to the rescue with some good pieces of advice.

According to the article, the introduction should clearly and succinctly provide the groundwork for the research to come. The backdrop for the research issue or problem should also be provided in the introduction, along with an explanation of the significance and value of the research. According to the author, the introduction should flow logically, with each segment building on the one before it. Beginning with a general review of the subject, the introduction should then gradually focus on the particular research question or issue that the dissertation will address. The article also suggests that, even if the dissertation is extremely technical or specialised, the opening should be written in a style that is understandable to a broad audience. The introduction should steer clear of jargon and technical phrases that readers outside the subject might not be familiar with, and it should give definitions for any specialised terms that are required. The author also proposes that a concise explanation of the research question or issue that the dissertation will address should be included at the conclusion of the introduction. This thesis statement needs to be succinct and explain what the research needs to do.

Overall, the piece emphasises the need to write a brief and clear introduction that demonstrates the significance of the research topic or problem, offers context and background information, and sets the stage for the research to come.

How to start a chapter

Commencing a chapter of the dissertation can present challenges, as there is a risk of inadvertently transforming the work into a lengthy journal article and structuring the chapter into excessively long sections.

---

[6] https://www.chronicle.com/article/on-the-dissertation-how-to-write-the-introduction/.
[7] https://www.chronicle.com/.

A dissertation is much more than that; it should tell a coherent story about your research. I find the advice proposed in the article 'In a Book or PhD, Start Each Chapter Cleanly, Never Link Back'[8] very valuable for students, since I tend to ask them myself to try to produce chapters that are sort of self-contained. Indeed, a chapter will be reviewed by examiners who will probably read your dissertation in bits, so if a chapter is self-contained without needing to go back and forth to find links back to sections, it will help the examiners to do their job properly and indeed lean them towards appreciating the work in its entirety. Adhering to this approach also facilitates the transformation of chapters into papers, enabling them to withstand the peer-review scrutiny of experts within the specific topic area and ultimately leading to the publication of research findings.

The article 'In a book or PhD, start each chapter cleanly, never link back'[9] examines the necessity of organising chapters in a book or thesis without constantly referring back to earlier portions. While it may be tempting to frequently refer back to prior sections to provide context and maintain a feeling of continuity, the author contends that this technique can actually impair the reader's capacity to grasp and recall information. Instead, the author recommends that each chapter be self-contained, with a clear and succinct introduction that gives background information. The author gives various reasons why this technique is advantageous. For starters, it helps the reader to concentrate on the content of each chapter without being distracted by references to earlier portions. This can improve the reading experience by making it more interesting and engaging. Furthermore, rather than frequently turning back and forth to prior sections, it can assist the reader in better absorbing the information by allowing them to build on their comprehension chapter by chapter. Finally, organising chapters in this manner might help to clarify and make the overall structure of the book or thesis more navigable.

Overall, the piece gives useful information about the significance of chapter organisation in academic writing. The author presents readers with practical tips that might help them develop more successful and entertaining books and theses by emphasising the necessity for self-contained chapters that prevent unnecessary linking back to earlier portions.

The author goes on to give some practical advice, which I believe is very useful:

- Beginning a new chapter neatly and with strong energy/high impact both marks the beginning of a new topic and urges readers to continue reading. An epigraph, a crucial statistic, a dramatic mini-case, a synoptic problem description, and a paradox are some components that can help achieve this.

[8] https://medium.com/advice-and-help-in-authoring-a-phd-or-non-fiction/in-a-book-or-phd-start-each-chapter-cleanly-never-link-back-3b2865e44173.
[9] https://medium.com/advice-and-help-in-authoring-a-phd-or-non-fiction/in-a-book-or-phd-start-each-chapter-cleanly-never-link-back-3b2865e44173.

- Following a high-energy start, the chapter's second component is some framing material that connects to the main body of the argument to come. The start component above is located and normalised in a properly drawn set of chapter borders by framing text.
- The third component of a chapter should start with a series of signposts pointing to the main body text sections. Signposts are informative but concise, and they are neither preludes nor mini-guidebooks to an upcoming argument. These cues should be quick and simple to grasp. If they are excessively long, the author should go back and decide where the main body content begins and then make a clean break with the first sub-heading. Authors can grab readers' attention and effectively launch new chapters by using these components: a high-energy opening, framing text, and signposts.

The writing anxiety

All the advice given before can be useful to overcome the daunting task of start writing a dissertation. Writing anxiety is a frequent sensation for many graduate students, and it can lead to procrastination, self-doubt, and poor thoughts about oneself as a writer. Writing anxiety can be induced by a number of circumstances, including peer pressure, fear of criticism, and a lack of confidence in one's own writing abilities. Setting realistic goals, breaking work down into manageable chunks, and building a regular writing practice are all ways to overcome writing anxiety. Seeking aid from peers, mentors, and writing centres can also be beneficial in overcoming writing anxiety.

To overcome writing anxiety, I would suggest to:

- *Create a clear outline:* before beginning the writing process, it is advisable to create a clear outline of the thesis or dissertation. This approach helps in organising thoughts and ensuring comprehensive coverage of all relevant themes.
- *Write every day:* the skill of writing can be refined through consistent practice. It is recommended to dedicate at least 30 minutes each day to writing. Establishing achievable goals and working towards them daily is essential in improving your writing abilities.
- *Break it down into simple pieces:* initially, writing a thesis or dissertation may appear daunting; however, breaking it down into smaller tasks can make it more manageable. Focus on completing one section or chapter at a time, and acknowledge your achievements as you progress.
- *Obtain feedback:* findings should be shared with advisors, peers, or other subject matter experts, as they can offer valuable criticism and aid in enhancing writing skills.

- *Edit and revise:* once the draft is completed, it is recommended to revisit and rewrite it. Reading the content aloud helps ensure smooth flow and coherence. Additionally, attention should be given to grammar, spelling, and citation accuracy.

Writing habits

PhD students should develop their own writing habits because each person has distinct skills, limitations, interests, and situations that influence their writing process. Developing excellent writing habits is critical for a successful academic career, and identifying the habits that work best for you can make the process more efficient, fun, and long-lasting. Furthermore, writing a PhD thesis or dissertation is a time-consuming and difficult procedure that demands discipline, focus, and patience. A routine supportive of productivity and progress can be established by cultivating individual writing habits. Techniques for managing writer's block, stress, and various demands on time and energy can also be attained. Discovering personal writing habits may enhance creativity, originality, and authorial voice. Experimentation with diverse writing approaches facilitates the identification of optimal methods, enabling the development of a writing style that reflects one's distinctive viewpoint and concepts. This process aids in distinguishing oneself within the field and producing compelling and impactful work.

Helen Sword's book *Air & light & time & space: How successful academics write* (Sword, 2017) investigates the habits and practices of effective academic writers. Sword discovers common tactics for overcoming writer's block, staying motivated, and completing high-quality research through interviews and surveys she conducted with over 100 academics from various disciplines. The book is structured into four sections, each of which focuses on a distinct aspect of the writing process: inspiration, structure, momentum, and polishing language. Sword provides useful advice and exercises for each stage of the process, emphasising the value of self-reflection and experimenting.

All in all, I consider that Sword's advice—sharing the stories and tactics of outstanding authors—can motivate PhD students to establish their own writing practices and succeed in overcoming the challenges they face while writing their dissertations.

Feedback on the dissertation

Frequent meetings with the supervisor will occur throughout the dissertation writing process, particularly upon completion of chapters or sections or when encountering difficulties in certain topics. It is recommended to adhere to the checklist below to ensure productive meetings with the supervisor and mutual agreement on subsequent steps:

1. Define the meeting's goals and agenda: Outlining the meeting's goals at the outset can foster progress. Establishing an agenda is essential to ensure that both the student and the supervisor are informed about the topics to be discussed and the tasks expected from them.

2. Set expectations: ensure awareness of the dissertation's requirements regarding scope, methodology, and deadline. Clarify any uncertainties by posing questions as needed.

3. Talk about writing techniques: the supervisor can be asked to share successful writing techniques and for the student to share their own. This may involve discussing strategies for structuring work, overcoming writer's block, and maintaining motivation during the writing process.

4. Review the writing process: the procedures for preparing a dissertation, including outlining, research, drafting, editing, and proofreading, should be discussed with the supervisor. Best practices for each phase can be reviewed, and guidance on staying on track can be sought.

5. Seek feedback: this may involve discussing research aims, theories, or plans and receiving constructive criticism and suggestions for improvement regarding the work.

6. Teamwork: inquire with the supervisor about initiatives among students such as subscribing to reading/writing groups or participating in peer-review sessions.

7. Meeting minutes: reiterate any key topics discussed and provide a meeting summary at the end. The summary should be shared via email, for example, and must be agreed upon by both parties: yourself and your supervisor. This way, during the next meeting, you can easily review the progress made or address any further questions or concerns.

The final examination

The format of doctoral examinations varies considerably across institutions and nations. Common approaches typically involve a faculty committee conducting a traditional oral examination, during which candidates are questioned about their research plan, expertise in the field, and related topics. Alternatively, examinations may entail a more comprehensive review of the field or a defence of the dissertation proposal. Some institutions may accept published papers, a portfolio of work, or significant research projects in place of or in addition to the oral examination. Candidates may also demonstrate their knowledge through published works instead of formal tests. Occasionally, institutions may assess candidates based on a substantial project or dissertation rather than oral examinations. The primary criteria for evaluating a candidate's suitability for a PhD typically include the quality, scope, and originality of their project. Finally, some universities or countries have shifted away from traditional PhD examinations and instead focus on evaluating candidates primarily based on their research achievements, coursework, and dissertations.

However, a final oral examination is traditionally part of a PhD defence or viva. A PhD viva, also known in Latin as a viva voce, 'by live voice', is a crucial part of the PhD examination process. It consists of an oral defence of the doctoral thesis or dissertation, where the PhD candidate must demonstrate their in-depth knowledge and understanding of their research.

The viva is conducted by a panel of examiners, typically consisting of an internal examiner from the candidate's institution and an external examiner who is an expert in the field. During the viva, the examiners will ask the candidates questions to probe their knowledge, test their ability to defend their work, and assess whether they have met the requirements for the PhD degree.

The viva usually lasts between 1.5 to 2 hours and is an opportunity for the candidate to elaborate on their research, clarify any issues, and demonstrate their research skills and contribution to the field.

Professor Emmanuel Tsekleves's LinkedIn post, 'PhDs, do you ever wish you could see inside an examiner's mind?'[10] presents a diagram of all the key questions examiners will typically ask in PhD vivas, as shown in Figure 3.1.

In my experience, too, as both an internal and external examiner, some viva examiner's questions can include: 'What are the strengths and weaknesses of your research?', 'How much of your work was original research, and how much was secondary research?, 'What original contribution has your thesis made to this field of study?', and 'What new thing do we know now as a result of your research?'

These types of questions are intended to evaluate the candidate's comprehension, expertise, and contribution to the subject of study, as well as their capacity to defend their research successfully in the viva.

Some useful references to prepare for the viva and to answer those type of questions can be found in a deck of slides published by Dr Ann M. Torres from the University of Galway, titled 'PhD viva guide—a springboard for your PhD viva preparation',[11] where various sections guide students to the different parts such as what to expect, how to prepare, the day of the viva, and after the viva.

---

[10] https://www.linkedin.com/posts/emmanueltsekleves_phds-do-you-ever-wish-you-could-see-inside-activity-7148667445891801088-EUY2/.

[11] https://www.universityofgalway.ie/media/graduatestudies/files/phdvivaguide/phd_viva_guide.pdf.

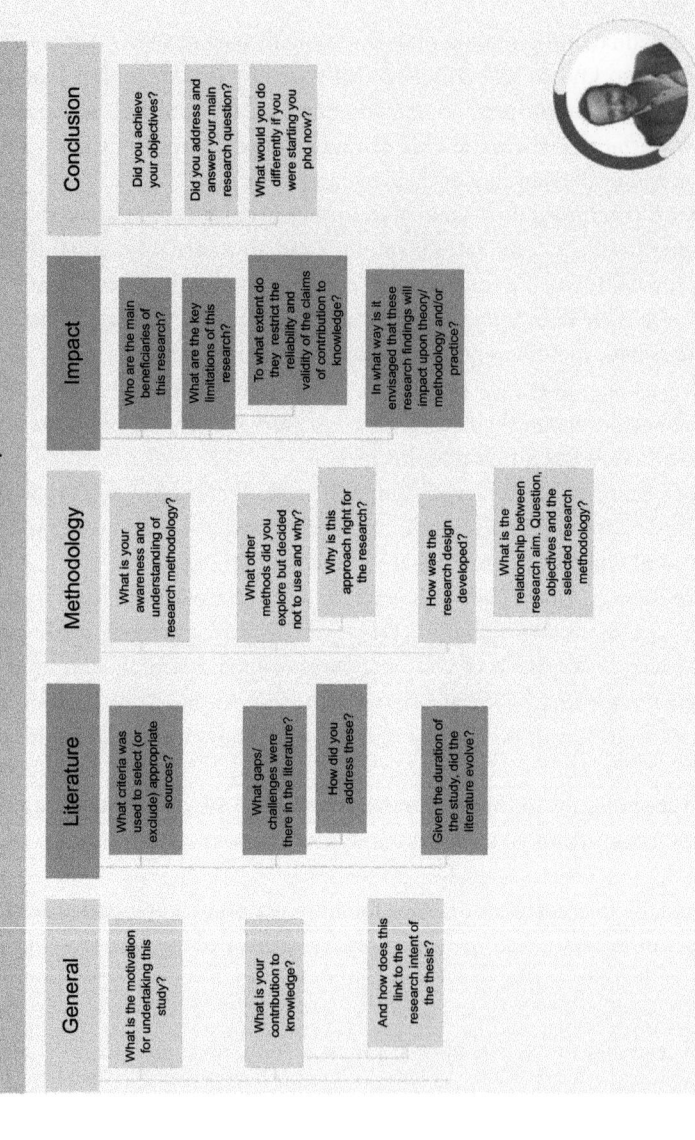

Figure 3.1 PhD defence, viva examiner questions

The article 'Top 12 potential PhD viva questions and how to answer them'[12] by Uttkarsha Bhosale focuses on preparing to answer questions during the viva or defence of a PhD. While each PhD viva is unique, there are certain commonalities in the kinds of questions that are asked. This article talks about the most typical and possible PhD viva questions, along with how to respond to them.

The possible outcomes of a PhD viva include passing without corrections, passing with minor or major corrections, or not passing at all. The examiners will provide feedback and recommendations based on the candidate's performance.

Preparation for the viva is crucial and can involve mock vivas with the supervisor, familiarising oneself with the examiners' work, and practising concise explanations of the thesis.

PhD students, depending on the institution, might be asked to give a brief presentation of their work at the beginning of the viva, normally with a duration of 5 to 10 minutes. Some institutions might require a more extensive presentation lasting from 30 to 45 minutes. Therefore, presentation skills are crucial at this stage, but also during the process of producing research to get feedback, for example, giving talks at conferences; presentation skills are discussed in the following section of this book.

Presentation skills

For PhD candidates, presentation skills are vital for various reasons:

- PhD students are expected to present their research findings to their colleagues, supervisors, examiners (e.g. viva), and, in some cases, at academic conferences. Effective presentation abilities enable the effective communication of research results to a selected or larger audience.
- Grant proposals: PhD students are frequently required to apply for research grants in order to fund their studies. Effective presenting abilities can assist the student in making a clear and convincing presentation of their research idea.
- Networking: other scholars in the field may be met at academic conferences and other activities. Students with strong presentation abilities can make a great impression and form vital connections.
- Presentation skills are very important for professional advancement. PhD students who can effectively explain their research findings will be more prepared to thrive in their chosen fields, whether in academia or industry.

[12] https://www.enago.com/academy/top-12-phd-viva-questions/.

Giving a killer presentation

The article 'How to give killer presentations'[13] (Anderson, 2013) offers tips on how to prepare and give excellent presentations. According to the article, presenters should focus on delivering an engaging story that interests the audience. The story should be clear, simple, and relevant to the interests of the audience. Anderson emphasises the need to practise the presentation in order to deliver the story successfully.

Anderson also recommends that presenters concentrate on the presentation's visuals. Visuals should support the story being told and be simple to understand. Anderson suggests using high-quality photos and graphics, and avoiding bullet points and elaborate diagrams. The presenter should also ensure that the visual aids are accessible to everyone in the audience, including those with visual impairments. The necessity of interacting with the audience is emphasised in the piece. Anderson recommends that presenters make eye contact, use natural movements, and speak enthusiastically. The presenter should also consider the audience's requirements and interests and modify the presentation accordingly. Anderson advises presenters to use questions, stories, and interactive components to engage the audience.

Finally, the article emphasises the importance of being genuine and passionate when giving a presentation. Anderson advises presenters to be themselves and let their enthusiasm for the subject shine through. This genuineness will contribute to the audience's trust and make the presentation more engaging. Anderson ends by reminding presenters that good presentations take practice and that anyone can become a killer presenter by following those principles.

While this article includes a very good piece of advice, I would like to add that a slide generally requires 1–2 minutes, so planning accordingly and rehearsing in front of a friendly audience who might give feedback is very useful. If an audience is unavailable for rehearsal, one can record the presentation and then review it with a positively critical attitude.

How to give an academic presentation

According to the suggestions in the article 'How to give an academic talk: Changing the culture of public speaking in the humanities' (Edwards, 1998), many academic talks are ineffectual, poorly organised, and lack participation, which can be harmful to the spread of research and ideas. Although the advice was given in the humanities, I believe that most of it applies to other areas in general, including information systems and computing.

[13] https://www.rphslibrary.org/uploads/1/1/0/9/110967909/public_speaking_readings.pdf.

A speech must accomplish three goals in order to be effective: presenting arguments and facts, convincing the audience of their validity, and being engaging and fascinating. Academics, on the other hand, tend to prioritise compelling argumentation over entertaining. Some even consider entertainment to be unimportant, believing that it detracts from the profundity of a presentation. This attitude is erroneous, since effective communication and persuasion are impossible without entertaining the listener. It is critical to entertain the audience, since their undivided attention is required to express the importance of the work being presented. Audiences at conferences frequently listen to multiple speeches over the course of several hours, making it difficult to retain focus. As a result, the speaker must make a concentrated effort to keep people interested. In an academic talk, entertainment does not always entail making people laugh or distracting them from their problems. Instead, it involves maintaining the audience's attention and interest throughout the presentation. Thus, entertainment in academic speeches is critical for keeping the audience engaged and focused on the subject being presented.

According to Edwards, any effective presentation must accomplish three goals:

(1) explain your ideas and facts,
(2) persuade your audience that they are correct, and
(3) be intriguing and entertaining.

Academics sometimes overlook the third point on this list in our fixation with persuasive argumentation. Sometimes, we believe it follows naturally from the first two. We even mock the goal itself at times. Especially in more technical/scientific areas, we seem to feel, perversely, that if a discussion is entertaining, it is probably not very profound.

The article goes on by suggesting the principles of effective talks (the three items mentioned above) and some useful rules of thumb: standing, moving, altering the pitch of your voice, speaking loudly and clearly towards the audience, creating eye contact, focusing on important ideas, using visual aids, and concluding the discussion within the allocated time are all advised ways to make a better academic talk. On the other hand, reading, sitting, standing motionless, speaking in a monotone, mumbling while facing downwards, staring at the podium, becoming lost in details, not having visual aids, running overtime, and not practising is usually worse. It is critical to summarise the essential points at the beginning and end of the presentation, to pay attention to the audience and adapt to its demands, and to copy exceptional presenters. It is equally critical to deliver a conclusion and observe the audience's behaviour.

Ensure that the presentation is concluded within the allocated time frame to prevent losing the interest of the audience. It's important to keep in mind that most people have a maximum attention span of 45 minutes, so going over that time may cause them to tune out, particularly during the conclusion, which is a critical part of a talk. Additionally, in a conference setting, exceeding allotted time can be seen as impolite, as it cuts into other speakers' time and discussion periods. Therefore, it's best to take responsibility for one's own time management and not rely on panel chairs to enforce the time limit. It's worth noting that finishing a few minutes early is always better than going overtime, which can create enemies and harm your reputation.

Here are some suggestions from the article for managing the timing of a presentation:

Firstly, if one uses PowerPoint or a similar presentation tool, one can establish a sense of timing by consistently using the same slide format. By giving a few presentations with the same format, one will become familiar with how long it takes to discuss each slide, allowing one to estimate the duration of their talk.

Additionally, when rehearsing for a presentation with a designated time limit, such as 20 minutes, it is helpful to annotate your notes at intervals of 5, 10, 15, and 18 minutes. This practice aids in time management and prevents exceeding the allotted presentation time.

Finally, it is important to refrain from improvising in front of an audience until one has fully mastered the art of presenting. However, techniques like humour or tangents can be employed to sustain the audience's engagement. Practising these aspects during rehearsal is crucial to gauge their timing accurately, akin to how a comedian or actor rehearses their performance.

To effectively communicate the main points and ensure that the audience remembers them, it is helpful to summarise the talk twice—once at the beginning and again at the end. I personally tend to give this advice to my PhD students even during the viva if they are allowed a brief presentation at the beginning. By following this guideline, as suggested in the article, the audience can grasp the key ideas and improve their retention of the message, which is ultimately the primary goal of a speaker.

A good technique to make your presentation more engaging and entertaining is to refer to the article 'For better presentations, start with a villain'[14] by Greg Stone (2015), published in *Harvard Business Review* in 2015. Stone claims that beginning a presentation with a villain can be more successful than beginning with a hero. Beginning with a villain, according to Stone, generates a sense of tension and conflict, which can help captivate the audience's attention and produce a more memorable experience. Stone recommends establishing a specific problem or challenge that the audience is facing and framing it in terms of a villain to use a villain in a presentation effectively. This could be a rival, a market trend, or a problem within the company that needs to be resolved. The audience is more likely to be emotionally invested in finding a solution if the problem is framed as a villain. Once, I presented a talk in the form of a murder mystery where the problem was

[14] https://hbr.org/2015/11/for-better-presentations-start-with-a-villain.

the murder and the hypothesis was based on cues, and I even got the best presenter award at that conference in computer science. Stone also emphasises the significance of having a villain and a hero in the story. While the villain instigates conflict and anxiety, the hero offers a solution and a sense of hope. The audience is more likely to be engaged and invested in the outcome if the presentation is framed as a conflict between the villain and the hero.

Overall, Stone's exploration suggests that introducing a villain at the beginning of a presentation can be an effective means of captivating an audience and generating a memorable experience. Presenters can generate a sense of tension and conflict that draws the audience in and inspires them to find a solution by pinpointing a specific problem or difficulty and framing it in terms of a villain.

Finally, in my experience, a simple and good way of structuring a presentation does not fall within the usual scheme for scientific presentations, especially if presenting at a conference where one might be given only a few minutes to give a talk (usually 10–15 minutes) but structure it as follows:

1. Start with a Twitter-friendly heading (brief) describing what the talk is about
2. Break down the key messages from the presentation into three separate slides (key message 1, key message 2, and key message 3)
3. Reinforce each of the key messages using either stories, statistics, examples, or case studies
4. If more information is required, break each major message into three subcategories that are reinforced in the same way.

In conclusion, to become an excellent speaker, it is recommended to learn from experienced and skilled speakers by observing their techniques and emulating them. It is important to pay attention to not only the words they say but also their body language, vocal inflexions, eye contact, timing, and handling of questions. By finding a good role model and putting in the effort to imitate them, one can become a successful speaker. For instance, listen to and emulate the great speakers at the 'How to give a great research talk'[15] event available on Microsoft.com on 24 July 2007.

Attending conferences and events

PhD students should attend conferences for a variety of reasons. For starters, conferences allow researchers to present their study to a larger audience, receive criticism and suggestions from professionals in the field, and earn publicity for

---

[15] https://www.microsoft.com/en-us/research/video/how-to-give-a-great-research-talk/.

their work. This can help them improve their research, raise their academic profile, and connect with possible colleagues.

Second, attending conferences enables PhD students to stay current on the latest advances and trends in their profession and learn about new techniques, methodologies, and technology. This can enhance their knowledge and comprehension of their field while also providing new views on their research effort.

Third, conferences provide opportunities for networking with other researchers, professors, and professionals in their field. This may result in new collaborations, job prospects, and exposure to fresh research ideas and methodologies.

Overall, attending conferences is a beneficial experience for PhD students since it allows them to present their work, keep up with current advancements, and network with other academics in their field.

Marta Teperek's blog 'How to make the most of an academic conference—a checklist for before, during, and after the meeting'[16] (Teperek, 2018) offers a detailed guide on maximising the benefits of attending an academic conference. The blog provides advice at each stage of the conference, beginning with preparation, when attendees should identify the sessions and talks they want to attend and plan their itinerary accordingly.

The article encourages guests to actively connect with the speakers and other attendees at the conference, asking questions, taking notes, and networking. During the conference, attendees should also take breaks and prioritise their health and well-being. Following the conference, the blog suggests following up with new connections created during the event, organising and sharing notes and resources, and commenting on the conference's insights. Attendees can make the most of their conference experience by following this checklist and developing lasting relationships and insights into their profession.

In particular, I would recommend that newcomers conduct preliminary research to identify the programmes, speakers, and other attendees in which they are interested and plan their itinerary appropriately. I would also recommend that PhD students use networking opportunities to meet new people, ask questions, and learn from previous attendees.

In conclusion, PhD students can also benefit from attending graduate consortiums (sometimes called doctoral colloquiums) or poster sessions in a variety of ways. For starters, graduate consortiums allow students to submit their research to a panel of experts in their field and obtain feedback and advice on how to enhance their research. This can assist students in refining their research topics, methodology, and approaches, as well as moving their research forward.

---

[16] https://blogs.lse.ac.uk/impactofsocialsciences/2018/03/16/how-to-make-the-most-of-an-academic-conference-a-checklist-for-before-during-and-after-the-meeting/.

Poster sessions allow students to present their research to a larger audience and engage in discussions with other researchers and educators in their field. This can assist them in increasing the visibility of their research, connecting with other researchers, and gaining new ideas and views on their work. Personally, I like to browse through poster sessions and ask questions to get an idea of new works in the field; indeed, while, as a scholar, I am generally aware of what other scholars are doing in the field, it is really from the poster sessions run by PhD students that novelty arises.

Attending graduate consortia or poster sessions can also help students improve their communication skills and practice presenting their findings clearly, concisely, and interestingly. This is a crucial talent that may be used in a variety of situations such as academic conferences, job interviews, and networking events.

Overall, visiting graduate consortia or poster sessions can be a great experience for a PhD student because it allows you to display your research work, receive feedback and guidance, network with other researchers, and improve your communication skills.

Dissemination

PhD students must disseminate their findings for a variety of reasons. For starters, disseminating research helps students enhance their profile and reputation within their field of study; students can enhance the visibility of their research and establish expertise in the field by publishing research in peer-reviewed journals, presenting at conferences, or disseminating findings on social media platforms. Indeed, sharing research is an important step in the advancement of scientific knowledge. PhD students contribute to the collective body of knowledge in their discipline and increase scientific understanding by sharing their results. This, in turn, can lead to new discoveries, improved treatments, and better technology.

Finally, research dissemination is an important aspect of the academic process. PhD students are required to share their results with their peers and contribute to the larger academic community as researchers. Disseminating knowledge is thus not only a chance for personal and professional advancement but also a responsibility to the larger scientific community.

Hopefully, the above reasons motivate you to understand that disseminating research is an important element of the PhD process, since it contributes to the advancement of scientific knowledge, the development of professional networks, and the establishment of the student as an authority in their field.

Prior to delving into the intricacies of scientific paper writing, it can be beneficial to examine potential venues that may serve as suitable targets for your research. While conferences, workshops, and, as suggested before, poster sessions and doctorate consortium meetings are all good starting points, there are also specific initiatives that focus on publications by PhD students. For instance, the Institute of Electrical and Electronics Engineers (IEEE) publishes *IEEE Potentials*, a journal for undergraduate and graduate students studying science, technology, engineering, and mathematics (STEM). Students can use the magazine to post articles, exchange research discoveries, and explore themes pertaining to their fields of study. *IEEE Potentials* can be a beneficial resource for PhD candidates for a variety of reasons. To begin with, the magazine serves as a venue for students to publish their research and discuss their discoveries with a larger audience. This can help you raise your profile and reputation in your field of study while also providing vital networking possibilities. Reading it can assist you in staying current on the newest trends and advancements in your field. The magazine covers a wide range of STEM issues such as developing technology, industry trends, and career advice. Students can learn about new research and advancements in their subject, as well as potential career pathways and possibilities, by reading the magazine.

Finally, *IEEE Potentials* might be a great resource for PhD candidates wishing to improve their communication and writing skills. By publishing articles in the magazine, students can refine their writing skills and acquire the ability to elucidate complex scientific concepts in a clear and concise manner. However, additional publication venues can be explored. It is advisable to seek recommendations from the advisory board regarding other comparable publication platforms dedicated to showcasing contributions from PhD candidates.

How to write an academic paper

In the realm of academia, crafting an academic paper as a PhD candidate may appear daunting at first glance. Nevertheless, through careful planning and meticulous execution, it is entirely plausible to generate a high-calibre paper that aligns with the established standards of your field. Guidance on planning can be found in a preceding section on planning[17] ('Planning to write a dissertation'), while this section is focused on enhancing PhD students' writing abilities in writing academic papers.

A general piece of advice is:

- Read carefully: To thoroughly understand the style and tone of academic writing, read widely and deeply in your discipline. Examine how other authors arrange their articles, use language, and credit their sources.
- Develop a clear and succinct writing style: Academic writing should be clear and concise; therefore, avoid jargon and unnecessarily complex phrases by

---

[17] https://docs.google.com/document/d/1MDIY99_g5f9OqaUuWMLiBPwUEEEiICU-ayYBgZnf2kE/edit#heading=h.hdusn5c9vdr5.

utilising plain language. When feasible, use an active voice and strive for clarity in your writing.

- Use a formal tone: Academic writing necessitates the use of a formal tone appropriate for the subject matter. Avoid using colloquialisms, slang, or abbreviations.
- Organise your thoughts: Good writing requires a well-organised mental process. Before you begin writing, make an outline of your paper, as suggested for your dissertation, to help you organise your thoughts and arguments.
- Use proper citing and references: Citing your sources correctly is an important element of academic writing. Make sure your references are thorough and exact, and use the appropriate citation format (APA, MLA, Harvard, etc.).
- Revise and edit your work: Revising and editing your work is an important aspect of the writing process. Examine your work for grammatical, spelling, and punctuation mistakes, as well as clarity and coherence of ideas.
- Obtain feedback: To receive comments and constructive critique, share your work with colleagues, professors, or a writing club. This will help you identify areas in which you may enhance and refine your writing abilities.

This is a checklist PhD students can always follow as general guidance; in the following, we will focus on some expert advice on scientific writing.

The presentation 'Scientific writing'[18] by Prof. Juan de Lara (2020), available on SlideShare, provides a comprehensive guide on how to write an effective scientific paper. The presentation is packed with good advice and emphasises the importance of clear communication and the need to adhere to certain conventions when writing scientific papers. The presentation is divided into several sections, with the first section providing an introduction to scientific writing and its importance. The second part delves into the various sections of a scientific paper, such as the abstract, introduction, methods, results, discussion, and conclusion, and provides a detailed overview of the purpose of each section and how it should be written. The final half of the slides focus on critical components of scientific writing such as adopting proper grammar and syntax, avoiding plagiarism, and comprehending the significance of referencing. The lecture also emphasises the significance of comprehending and adhering to the rules of the publication to which the manuscript is being submitted.

More advice I find useful for PhD students comes from the paper 'How to write an academic paper'[19] (Ringmar, 2015); it gives detailed instructions on how to produce a fruitful academic paper. Ringmar makes the argument that creating an academic paper is a skill that can be developed via practice rather than a mysterious procedure. He urges authors to stay away from making generalisations and

---

[18] https://www.slideshare.net/slideshow/scientific-writing-235422422/235422422.
[19] https://lucris.lub.lu.se/ws/files/6359979/8165107.pdf.

emphasises the significance of having a clear, short thesis statement. Additionally, he advises writers to follow a logical structure and provide solid proof to back up their claims. The paper stresses the significance of research in academic writing. It suggests that authors begin by conducting a thorough literature review to get a firm grasp of the subject. They should then concentrate on a smaller area and obtain knowledge that is more specialised. Ringmar advises authors to assess their sources cautiously, taking into account their applicability and dependability.

Finally, Ringmar provides helpful advice on how to draft and edit an academic work. He suggests that authors begin with a rough draft and then go back and edit it several times. He advises authors to read their work aloud to identify faults and ensure that it flows naturally. He also advises authors to check their work for grammatical and spelling errors carefully. Overall, Ringmar's piece offers PhD students a helpful foundation for creating an academic paper that is thorough in its research, rationally organised, and backed up by solid evidence.

Good guidance and a series of tips on writing first-class academic papers can be found in the article 'How to write a first-class paper', where Gewin (2018) gives instructions on how to create a top-notch research paper that stands out in the sector. The article suggests that researchers start by meticulously preparing their study and specifying the format of their report. Before beginning to write, it is essential to have a firm grasp of the research topic, the technique, and the findings. The document should also be logically organised, with a succinct introduction, a part on techniques, a section on results, and a section on comments.

The paper emphasises the need for clear and succinct writing and claims that effective writing can significantly affect how a paper is perceived. The text should be simple to understand, brief, and clear. The authors should avoid using obscure technical words and jargon. Instead, they ought to explain whatever technical jargon they employ and speak plainly. For the work to be credible and professional, correctness, spelling, and grammar must also be carefully considered.

Additionally, Gewin offers detailed advice for each part of a research paper. The major conclusions of the work should, for instance, be succinctly and plainly summarised in the abstract. The research topic, methods, findings, and conclusion should all be included. The research question and its importance should be summarised in the introduction. Additionally, a short and unambiguous thesis statement should be included. The results part should describe the findings in a clear and succinct manner, while the methods section should be sufficiently detailed to enable another researcher to duplicate the work.

The article concludes by highlighting the significance of editing and reworking the paper. PhD students should ask their peers and fellow colleagues for input to find out where they might improve. In order to make sure the document is error-free and flows naturally, it is also crucial to proofread it properly. In conclusion, the article exhorts researchers and PhD students to take pleasure in their work and make an effort to produce a study that significantly advances their discipline.

These recommendations will help you create excellent papers that might have a long-lasting effect on their respective field of study.

Josh Bernoff's book *Writing without bullshit* (2016) gives another good piece of advice that attempts to help professionals communicate more effectively by cutting through the clutter and getting right to the message. Bernoff contends that using jargon, buzzwords, and intricate language not only impedes communication but also destroys credibility and wastes time. He argues for a 'bullshit-free' writing style that emphasises clarity, brevity, and genuineness.

Bernoff provides practical strategies and techniques for writers of all levels to do this, from recognising and deleting extraneous words and phrases to generating effective titles and shorter sections. He also emphasises the significance of knowing the audience and adapting the message to their needs and interests—for instance, an academic audience—but the message might still vary depending on the venue you would like to address. I believe that PhD students can enhance their communication skills, gain trust with reviewers and peers, and eventually progress in their careers by following Bernoff's guidance.

Lennart E. Nacke's paper 'How to write and review CHI papers' (2017) explains how to prepare and assess research papers for the CHI (computer-human interaction) conference. Although it was written for such a conference, the advice is general enough to be useful in various fields. According to Nacke, the objective of papers is to disseminate new research findings to the community; therefore, they must be clear, short, and well organised. The guidance is general enough to be useful for writing an academic paper; in fact, it describes the structure of a paper, which includes an introduction, related work, methods, results, and conclusion. It is necessary to properly define research objectives and hypotheses, provide extensive information about procedures and participants, and present results in a clear and succinct manner using a ROAM analysis proposed by Bernoff (2016). A ROAM analysis is a useful exercise to undertake before writing a paper, and it consists of four phases: R (readers) is scoping clearly who the audience is; O (objectives) how to address the readers; A (action) what would the readers learn from the paper, and M (impression) what is the main takeaway message of the paper. Nacke's advice is also to consider the following questions:

1. What constitutes the problem that needs to be solved?
2. Why is it important to solve this problem?
3. What solution has been identified to solve it?
4. How do we know that the solution is a good solution to the problem?

Finally, an extended abstract published at CHI (The ACM CHI Conference on Human Factors in Computing Systems) by Nacke (2018) revealed a key message from senior CHI members, and again, it holds in general: when writing an academic paper, focus on correctness, validity, and a contribution that you believe might change or progress the field significantly.

The course went on to describe how to structure an academic paper, which I always use as a template for my students:

1. Define a research question
2. Test your questions against the literature
3. Refine your research question
4. Derive a method that answers your question
5. Plan out a detailed analysis
6. Refine your method and analysis
7. Execute your method
8. Execute your analysis (this is your Results draft)
9. Write your discussion
10. Polish! Polish! Polish!
11. Submit and wait
12. Do it all over again!

Another methodology useful to structure an academic paper, but I would add a dissertation too, presented in that course at ACM CHI 2017, was the hourglass structure (Rana, 2022); see Figure 3.2.

1. Abstract
2. Introduction
3. Literature review
4. Methodology
5. Results
6. Discussion
7. Conclusion
8. Acknowledgements
9. References

Once the paper is written, it will need editing for clarity. Clarity editing is a crucial step in finishing an academic work after it has been completed. In academic writing, clarity is essential since it helps your readers understand your ideas. Start by going through the paper's broad structure and organisation to improve clarity. Make sure the introduction offers a clear thesis statement that establishes the course of the argument and that the body paragraphs flow logically from one to the other. Make sure that each paragraph contributes to the overall coherence of the paper, and revise and reorganise sections as appropriate.

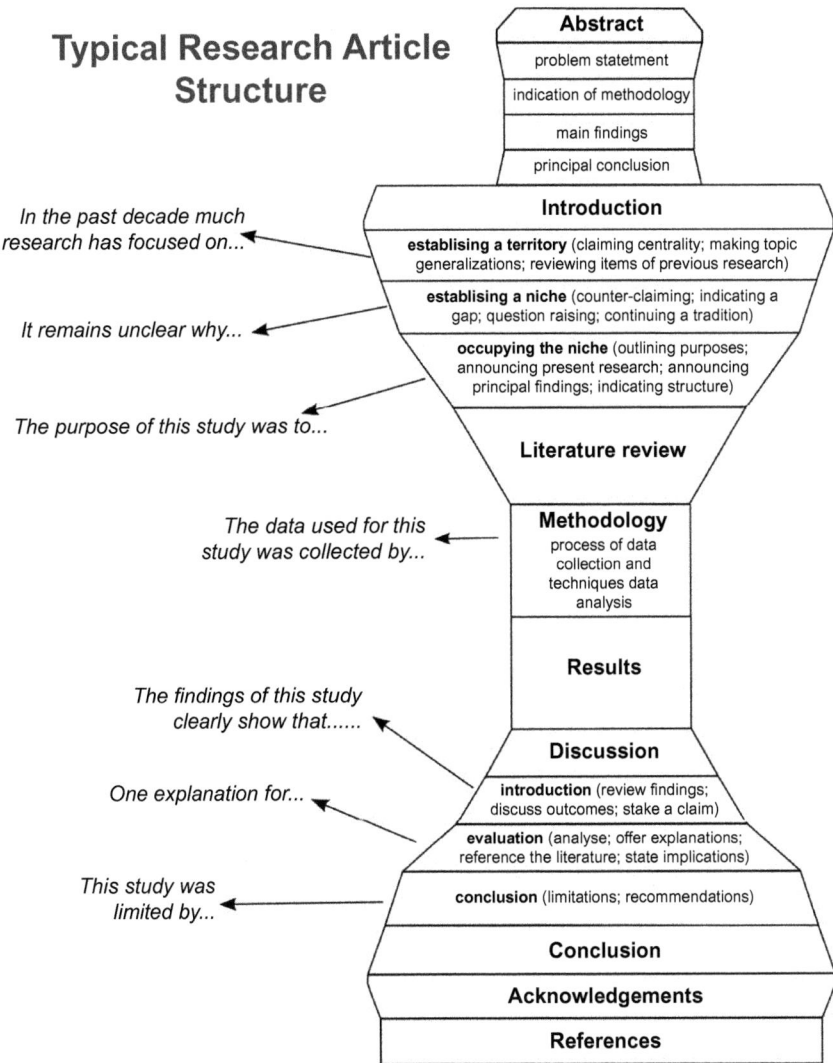

Figure 3.2 Typical research paper structure.

Source: https://www.researchgate.net/publication/359815719_Typical_Research_Article_Structure

Next, concentrate on streamlining your sentences and getting rid of any unclear or confusing terminology. Avoid using superfluous jargon or overly complex vocabulary that can mislead readers by speaking in a straightforward and succinct manner. To make the writing more direct and interesting, think about employing an active voice wherever possible. In addition, pay attention to how well the paragraphs flow into one another. Make sure each paragraph has a distinct theme sentence and that the following sentences build on and support it.

Thoroughly check the paper for potential grammatical flaws, typos, punctuation faults, and structural and linguistic changes. Check the writing carefully for subject-verb agreement, consistency of tense, and appropriate punctuation. Grammar checkers and other tools can be useful, but you should always rely on your own judgment because they might not capture all faults or recommend the best fixes.

It is also beneficial to seek input from others, such as peers or instructors, since they can offer new insights and point out areas that might need more explanation. It is important to bear in mind that while you ultimately have the authority to decide on revisions, it is advisable to consider the advice provided to you.

Finally, spend some time reading your essay out loud. This method can assist in locating problematic wording, ambiguous sentences, or any other elements that might sound perplexing. Reading one's work aloud can assist in accurately assessing the flow of writing and identifying areas that may require further editing.

Always remember that perfecting academic work requires time and careful attention to detail. Processes such as editing for clarity, restructuring sentences, verifying proper syntax and punctuation, soliciting feedback, and engaging in oral reading of the text can significantly enhance the quality of academic writing.

A set of questions that can guide the revision of a paper stems from the course at ACM CHI 2017, which here is grouped into topics in Table 3.3:

Table 3.3 Questions to ask to revise a paper before submission

Clarity aspect	Questions to consider
Vocabulary	Do I use simple, familiar words throughout?
Sentences structure	Are any sentences too long or contain too many sub-clauses?
Jargon and expression	Have I eliminated all jargon and clichés, hedging?
Active voice	Do I use the active voice?
Sentence construction	Do I avoid reversing into sentences?
Tense consistency	Are my tenses consistent?
Word usage	Do I avoid word echoes?
Logical thinking	Does my writing reflect clear, logical thinking?
Coherence and readability	Will a reader be forced to reread anything?
Eliminating clutter	Have I removed all the unnecessary clutter?
Conciseness and directness	Have I got to the point immediately?
Repetition and redundancy	Have I avoided repeating words or points?
Excessive detail	Have I gone into too much detail anywhere?
Ambiguity	Have I avoided ambiguity?
Simplification	Do I avoid making simple points sound complicated?
Distinctions and comparisons	Are all my distinctions/comparisons crisp?
Word trimming	Can I cut out any more words?

Box 3.1

I found the advice given by Dr Anna Clemens, an academic writing coach, in a thread on X,* to be very useful for reflecting on the typical mistakes supervisors can make when co-writing papers with PhD students. I use this piece of advice almost as a checklist to remind me of the steps to be taken when collaborating with PhD students in drafting papers.

The typical mistakes mentioned by Dr Clemens are the following:

1. Not teaching the steps of the writing process
2. Only editing for language, not structure
3. Not defining what a story is
4. Not setting internal deadlines
5. Not providing reasons for your edits
6. Having your student write a complete draft first thing
7. Thinking reading the literature teaches writing
8. Not checking in with your students about what they may need

The thread includes a brief unpacking of those steps, which I recommend reading. Based on my experience, this list of steps is useful, particularly for early career supervisors, to circumvent common pitfalls when engaging in collaborative writing activities with PhD students.

 * https://x.com/scientistswrite/status/1679178890529603584?s=12&t=8Fd0cSO7d7ADtK2wIXB1sg&mx=2.

How to get published in an academic journal

Academic journal writing is very competitive, as stated in the article[20] that inspired the very title of this section. It is even more competitive in 2025, as described by Kenfra Research, a service provider that specialises in providing PhD assistance, in a similar article entitled: 'Finding the right PhD journal publication in 2025: A comprehensive guide'.[21]

Even upon overcoming the initial challenge of generating a valuable topic or research piece, the subsequent task of summarising it in a manner that captures the interest of reviewers remains.

[20] https://www.theguardian.com/education/2015/jan/03/how-to-get-published-in-an-academic-journal-top-tips-from-editors?CMP=share_btn_tw.
[21] https://kenfra.in/finding-the-right-phd-journal-publication-in-2025-a-comprehensive-guide/.

There is no easy formula for publication; editors' standards can differ between and even within subject areas. Regardless of their discipline, PhD students will face some difficulties. In response to reviewer criticism, it is essential to address each point systematically, providing clear explanations or revisions where necessary. Adhering to proper paper formatting guidelines ensures consistency and professionalism in academic writing. Continuous editing and resubmission may be necessary to refine the paper and address reviewer feedback comprehensively.

Summarising the advice given in the article on submitting an academic paper:

1) Select the appropriate journal:
   - Ensure that your article aligns with the scope of the journal.
   - Familiarise yourself with the editorial board and recognise some of the members.
   - Review recent issues to gauge the journal's focus, quality, and impact.
2) Follow submission procedures:
   - Carefully read and adhere to the instructions provided by the journal for authors.
   - Spending a few minutes on this step saves time for both you and the editor.
3) Craft an effective cover letter:
   - Use the cover letter to highlight the most interesting and significant aspects of your paper.
   - Avoid repeating the abstract or providing a detailed summary of the paper.
   - Focus on providing a broader outline and any additional relevant information.
4) Provide context for your research:
   - Clearly demonstrate where your research fits within the broader scholarly landscape.
   - Address the gaps in knowledge that your research aims to fill.
   - Lack of context or clarity about the research's importance can lead to rejections.
5) Accurately describe methodology:
   - Be honest and specific about your research methodology.
   - Avoid overstating or misrepresenting the scope or depth of your data collection.
   - Use theory consistently throughout your argument and text if it supports your analysis.

Tips for dealing with feedback:

6) Respond to reviewer comments:

- When revising and resubmitting, summarise all changes suggested by the reviewers.
- Remain objective and avoid venting frustrations.
- Take time to process the feedback, discuss it with others, and draft a thoughtful response.

7) Revise and resubmit:
   - Don't give up after receiving a 'revise and resubmit' decision.
   - Persevere through the revision process, even if major changes are required.
   - Some authors who tackle major revisions end up successfully publishing their work.

8) Justify challenging reviewer suggestions:
   - It's acceptable to decline reviewer suggestions if you can provide a good justification.
   - Engage in a polite and rational discussion to present your counterarguments.
   - Show that you have considered the feedback and incorporated relevant suggestions.

9) Assess publication speed:
   - Understand that certain journals may take longer to publish than others.
   - Determine the urgency of publishing your work based on your needs and goals.
   - Some journals offer advanced access, which can be advantageous for specific circumstances.

10) Recognise the publication process:
    - Published papers only show the final version, not the earlier drafts or failures.
    - Remember that publishing in top journals is challenging for everyone.
    - Do not be discouraged by the polished works you encounter; everyone goes through revisions.

By following these tips, PhD students can improve their chances of successfully submitting and publishing academic papers.

Publishing academic papers is essential for PhD students because it enables them to add to the body of information already known about their subjects and helps them establish a research profile. Students who share the results of their research not only improve their knowledge of the subject as a whole but also identify themselves as capable researchers. The ability to perform rigorous and significant research is demonstrated by publishing articles, which is essential for developing a solid research profile. Additionally, it raises the visibility and acceptance of their work within the academic world, offering doors for partnerships, networking chances, and potential future careers.

Furthermore, it is crucial for establishing academic credentials and career growth. Academic paper publication is a crucial part of the professional growth of PhD students. It serves as concrete proof of their intellectual accomplishments, demonstrating their know-how to prospective jobs, academic institutions, and funding organisations. In order to establish a solid academic reputation, which is important when applying for academic positions, research funding, and fellowships, students must publish papers. It also makes them more credible as subject matter experts, allowing them to take part in intellectual gatherings like conferences and symposiums and promote their careers even further.

For PhD students, writing and publishing academic articles is a transformational experience. Developing critical thinking, analytical abilities, and effective communication of study findings promotes intellectual growth. Students can get helpful criticism through peer review and revisions, which helps them to improve their approaches and arguments. Students gain resiliency, flexibility, and a greater comprehension of their research field through this iterative approach. They can improve their research topics, test their theories, and participate in ongoing scholarly conversations by publishing articles, which encourage both academic and personal development.

What happens after submitting a paper

After submitting a paper to a conference, there are multiple possible outcomes. Here are some typical examples, but the precise results can vary based on the conference and its review procedure:

- Acceptance: The paper might be approved for conference presentation. This indicates that the conference planners and reviewers thought the paper was well written and pertinent to the conference's theme. At academic conferences, presenters are typically required to deliver their papers either orally or in the form of a poster presentation. The paper will very likely be published in the conference proceedings.
- Conditional acceptance: When a work is conditionally accepted, it means that the reviewers have questions or ideas for revisions. In such circumstances, the conference organisers could provide feedback or ask for revisions. Responding to these issues and submitting a new draft of the paper for final approval will be possible.
- Rejection: Regrettably, if the reviewers decide that the paper is inappropriate for the conference, they may reject it. Rejection can happen for a number of different reasons, including a lack of innovation, poor writing, a lack of research contribution, or a mismatch with the conference's aim. Rejections are frequent; therefore, it's crucial to take the criticism positively and think

about rewriting the article before submitting it to a different conference or publication.

Apart from acceptance or rejection, there are other possible outcomes when a paper is submitted to a journal:

- Minor revisions: The reviewers might find the paper promising in some circumstances, but they might also have a few minor questions or ideas for enhancements. Before the paper is approved for publication, the journal can ask the author/s to submit these amendments. The paper will probably be accepted if the author/s responds to the reviewers' suggestions and makes the required adjustments.
- Major revisions: The reviewers may advise major revisions if they find significant problems with the research such as methodological errors, shaky analysis, or gaps in the literature review. In this scenario, thorough criticism and recommendations for enhancing the article will be provided. The paper will need to be revised in order to address the reviewers' issues. The paper will go through another round of reviews following the amendments to see whether it has been ultimately accepted.
- Reject and resubmit: A paper could occasionally be rejected, along with a request to resubmit. This indicates that even while the reviewers think your work has promise, they think important changes must be made before it can be accepted. The journal might offer detailed criticism and advice on editing and resubmitting the manuscript. In such circumstances, the opportunity to address the reviewers' criticism and enhance the article before resubmission is available.

A conference paper is submitted 'as-is'. Even if the authors can make a few corrections to the work before it is ready for publication, the reviewers (at least the externals) have no means of knowing if this has been done.

Authors frequently have the chance to submit a response after reading the reviewers' criticisms. In a written document called the response, authors can address each criticism given by reviewers. Authors may defend their arguments, clarify any misunderstandings or misinterpretations, offer more data or analysis, or make changes in response to reviewers' suggestions in the rebuttal. It is preferable to approach writing a response with the attitude that 'our paper already has all the information you are asking for; it is just a matter of looking'. Although it may not be strictly correct, a significant flaw cannot be acknowledged in a rebuttal because doing so will surely result in a rejection (and if there is a significant flaw, you probably should not be submitting a rebuttal in the first place). Keep this tone throughout the entire argument. Writing an effective response is difficult because there is a delicate line between sounding haughty and not being.

Concerning rebuttals, I found good advice in Niklas Elmquist's (2016) article 'Writing rebuttals', which provides insightful advice on creating strong rebuttals in academic writing. Rebuttals are important in the peer-review process, and the author underscores their possible influence on the conclusion made about a submitted manuscript.

Elmquist starts by discussing how crucial it is to comprehend and analyse reviewers' comments. He recommends that authors approach criticism with an open mind, refrain from getting defensive, and try to pinpoint the key issues and ideas made. The author stresses the need for respectful, professional responses to each criticism, appreciating the reviewers' knowledge and properly addressing their concerns. The article offers a step-by-step tutorial on how to organise a response. Elmquist advises beginning with a succinct introduction that thanks the reviewers for their time and work and acknowledges them. The author advises employing a clear and succinct writing style to guarantee that the message is successfully communicated.

Elmquist then suggests responding to each criticism separately and structuring the refutation by the reviewers' comments. The author proposes responding to each criticism with a concise assessment of the issue raised and a detailed explanation of how it will be resolved. Elmquist highlights the significance of supplying proof or logical justification to back the response, strengthening the author's argument. The author also emphasises the importance of keeping a polite and authoritative demeanour during the reply. Elmquist cautions against using harsh or combative language since it can undermine the authority of the author's claims. Instead, the author advises concentrating on the academic or scientific components of the topic and offering impartial justification.

The article also emphasises the significance of editing the original work in light of reviewers' suggestions. Elmquist advises viewing the reviewers' criticism as a chance for progress and stresses the significance of explicitly outlining the changes made in response to their advice.

The importance of crafting strong rebuttals is emphasised in the article's conclusion, since they can have a big impact on how the peer-review process turns out. Elmquist advises authors to view rebuttals as a chance for mutual understanding and a constructive debate that will ultimately raise the calibre of their submitted paper.

How to review a research paper

PhD students have the opportunity to develop their analytical and critical thinking abilities by participating in the peer-review process. Reading papers can help them gain the skills necessary to evaluate the pros and drawbacks of research, spot knowledge gaps, and provide helpful criticism. These abilities depend on their own research and future academic careers.

PhD students are exposed to a variety of research subjects and approaches through peer evaluation. It gives them a chance to stay current with new developments in their profession and gain an understanding of various methods and viewpoints. Students can broaden their knowledge and gain a more comprehensive picture of the academic environment by reviewing papers.

Furthermore, PhD students get the opportunity to communicate with seasoned academics in their field through peer reviewing. They can create connections, relationships, and collaborations through the review process. This networking can result in chances for mentorship, joint publications, and future partnerships, all of which are essential for career growth.

It is important to give authors of reviewed papers precise, straightforward feedback. This method, by critically analysing the organisation, coherence, and clarity of others' work, will aid PhD students in developing their own writing and communication abilities. Additionally, it enables them to improve their capacity for efficient verbal and written communication.

Peer review is an academic community service. PhD students improve the quality and reliability of published research by offering meaningful and helpful criticism. By reviewing papers, they can actively contribute to the scholarly conversation, influence the development of their subject, and assure the dissemination of rigorous and significant research.

Finally, peer-review participation shows dedication to academic rigour and capacity for critical analysis of research. By actively participating in the peer-review process, one can gain more respect and credibility in the academic world. Additionally, it provides a vital experience that PhD students may showcase at job interviews and on their resumes.

In his article 'How to constructively review a research paper,'[22] Narayanan offers insightful advice on how to write reviews of research papers that are both helpful and productive. The author stresses the value of the peer-review procedure in preserving the quality and reliability of scholarly investigation. Before assessing a manuscript, Narayanan emphasises the importance of being conversant with the study field. To offer thoughtful comments, he advises reading relevant publications and being familiar with the body of previous research. The author suggests that reviewers approach the work with optimism and concentrate on the paper's advantages and prospective contributions rather than just looking for problems.

The piece offers a step-by-step tutorial for putting together a helpful review. Narayanan proposes to introduce the paper's core contributions and major conclusions with a succinct overview. The novelty and importance of the paper within

---

[22] https://freedom-to-tinker.com/2018/05/15/how-to-constructively-review-a-research-paper/.

the study field should be highlighted in this concise overview. The author then suggests offering a fair assessment of the paper's advantages and disadvantages. Reviewers are urged by Narayanan to be detailed in their remarks and to back up their judgments with examples and supporting data. It is important to provide constructive criticism in a respectful and appropriate manner, concentrating on the enhancement of the work rather than criticising the writers.

Narayanan also stresses the value of making concrete recommendations for development. Reviewers should provide detailed suggestions on how the authors might improve the paper's weaknesses or build on its strengths. To substantiate these proposals, the author advises giving precise justifications and citations to previous works of literature.

Summarising, Narayanan's article offers a thorough manual on providing constructive criticism during the peer-review process. The paper stresses the value of expertise in the research field, an optimistic outlook, and a fair assessment. It emphasises the merit of detailed and helpful criticism, which enables writers to enhance their articles and expand scientific understanding. Reviewers can effectively play their part in preserving the calibre and integrity of scholarly research by adhering to these rules.

Beyond traditional metrics

The evaluation of scholarly research has historically been linked to the application of conventional measures, like the H-index and journal impact factors (JIF). A scholar's H-index is calculated based on the number of papers (H) cited at least H times, while the number of citations to a paper in a journal determines the JIF, which is frequently used to evaluate the reputation of journals and, inadvertently, the quality of work published in them. For many years, scholars have depended on these metrics to assess the quality and importance of their research, which has enabled them to obtain funding, publish in esteemed venues, and advance their careers in academia. Nonetheless, the scholarly discourse has recently focused more on the shortcomings and biases included in these conventional measurements. We are witnessing a transformation in the field of research assessment as it moves away from standard metrics towards a more transparent, inclusive, and holistic approach that places significant emphasis, for example, on the impact—intended as the practical consequences and uses of academic research. Such impact can be evaluated in a number of ways such as case studies, narratives, and proof of how scholarly writing and research have influenced developments, enhancements, or breakthroughs in many sectors.

Conventional measures such as the H-index and the JIF have been useful substitutes for evaluating the influence and quality of research. These measurements do, however, have a lot of shortcomings. The JIF, for instance, may encourage journals to place more emphasis on timeliness and sensationalism than on the quality of the research they publish. Additionally, it may hurt publishers that produce books or other long-form content that may not immediately gain a lot of citations such as journals in developing fields. Although an effective single measure, the H-index concentrates on a small number of papers and might not fully represent the range of a researcher's contributions.

The environment for research assessment is changing as a result of these limitations. The growing acknowledgement of books as a significant type of scholarly work is one noteworthy change. Books are especially vital for fields where long-form, thorough study is essential, since they provide a forum for in-depth idea investigation and synthesis. Books are fundamentally interdisciplinary, frequently requiring teamwork, and can play a significant role in addressing societal challenges. Additionally, they play a crucial role in bridging the knowledge gap between the general public and academia by making knowledge available to a wider audience. As a result, a growing number of institutions and funders are now taking books into account in research evaluation processes in addition to standard journal articles and conference papers. Books are relevant because they inform and involve participants, keep track of common methods and outcomes, and promote the benefits of citizen science. They eventually increase the effectiveness of citizen scientific efforts by acting as important resources for data analysis, community development, and the public sharing of research findings. Due to its capacity to involve the public, collect a variety of data, and increase public knowledge of scientific issues—all of which contribute to better decision-making and workable solutions for real-world problems—citizen science is a good approach for assessing the social impact of research. Involving citizens in research empowers communities, builds public trust, and establishes a direct link between research findings and changes in policy and society. The wide term 'citizen science' refers to the active involvement of members of the public in scientific research projects (Vohland et al., 2021). The goal of this cooperative approach between scientists and the general public is to produce new information that will advance both science and society. Although citizen science as a discipline has centuries-old roots, the term itself was established in the 1990s and has since been widely recognised. Recognition of citizen science is growing in a number of fields such as science, policy, education, and society at large. It is becoming a recognised area of study and application, with a focus on the need for a thorough understanding of terminology, standards, and norms.

Alternative evaluation models that go beyond conventional criteria are being investigated by funding organisations and research institutes. These strategies emphasise open science, interdisciplinary cooperation, and public participation while focusing on the quality and societal impact of research. The problems with the JIF are addressed by new metrics such as the relative citation ratio (RCR), which takes into account an article's influence in relation to its field. Bloudoff-Indelicato (2015) states that the RCR is determined by dividing the total number of citations a paper receives by the average amount of citations an article in that subject receives. After that, the figure is compared to the median RCR of all papers published in the same area. This makes it possible to evaluate articles according to their significance within their own fields, and highly influential works will be acknowledged regardless of whether they are published in a niche publication. Finding the average number of citations for every other article that is cited alongside the article for which the RCR is being calculated (also known as 'an article's co-citation network') yields 'the average number of citations an article usually receives in that field'.

Furthermore, the use of narrative assessments is growing in popularity, in which researchers give their work context and significance. Using this method enables researchers to discuss the importance of their contributions and how they fit with the larger objectives of their research. Traditional and open peer reviews are being used to offer a qualitative assessment of research. For example, the UK's method for evaluating the quality and significance of research carried out at higher education institutions is called the research excellence framework (REF). It entails a thorough and rigorous assessment of the results of the research, the context in which it is conducted, and the social and financial implications of that study. Government financing to universities is determined by expert panels evaluating the quality and importance of research in a range of fields. Approximately every six to seven years, REF assessments are conducted. They are an essential instrument for funding allocation, accountability, and fostering research excellence in UK academic institutions.

The way that research assessment is currently being conducted is evolving from the past when measures like the H-index and JIFs were used. This denotes a change in the direction of a clearer, nuanced, and inclusive appraisal of scholarly contributions.

However, managing this shift calls for coordinated efforts from a number of stakeholders, including institutions, funding agencies, governments, and academics. Collaborative endeavours aimed at developing robust evaluation frameworks that consider qualitative, quantitative, and societal impacts are crucial; for example, the coalition for advancing research assessment (CoARA) initiative.[23] Enough resources and support systems must be made available to researchers so they can pursue unorthodox ideas without worrying about their work being negatively impacted by evaluation procedures.

---

[23] https://coara.eu.

Open science

The ideas of open science have acquired considerable support in the ever-changing field of academic research. They have changed scholarly procedures and placed a strong emphasis on transparency, inclusivity, and knowledge sharing. Adopting the ideas of open science not only helps PhD students stay in line with ethical research procedures but also encourages innovation, speeds up discoveries, and increases the impact of their work. Open science, which encompasses a range of methods, represents a paradigm change in scholarly communication (Miedema, 2022).

Open access (OA) is a new method that is transforming the landscape of scholarly communication by making scholarly material freely available to everybody. This paradigm guarantees unlimited access to scholarly articles, research data, and instructional tools, challenging the conventional subscription-based model of access to academic publications.

Open access is based on a number of fundamental ideas (Piwowar et al., 2018). Removing paywalls and subscription costs removes obstacles and makes academic literature, research findings, and instructional materials more accessible. This inclusiveness overcomes financial or geographic barriers to promote global engagement and information dissemination. OA also advances equity by democratising information access, opening up scholarly knowledge to a wider range of groups, encouraging interdisciplinary collaborations, and advancing societal achievements. Additionally, OA greatly increases the visibility and potential effect of research by removing obstacles to access, which speeds up the use of knowledge in both academic and non-academic contexts (Langham-Putrow et al., 2021).

There are two main open-access models. Authors who publish in open-access journals can share their articles freely as soon as they are published, a practice known as gold OA. These journals frequently use article processing charges (APCs) to defray costs. Conversely, Green OA, often known as self-archiving, entails authors storing a manuscript version in repositories (e.g. arxiv[24]), making it freely accessible following an embargo or acceptance.

OA has many benefits. It promotes cross-border knowledge exchange by facilitating global knowledge sharing and providing academics, students, and practitioners with access to important information. Furthermore, OA promotes scientific advancement and innovation by accelerating the dissemination of research findings, resulting in the quicker translation of discoveries into real-world applications. Furthermore, works that are freely accessible frequently receive more citations, which raises the visibility, recognition, and possibility of collaborations for study.

---

[24] https://arxiv.org/.

Even with its advantages, OA has drawbacks. Its survival depends on maintaining viable funding models such as transformative agreements or public funding support. Sustaining research rigour and integrity in open-access journals and archives requires the maintenance of strong peer-review procedures. Furthermore, managing licensing and copyright agreements is essential to protecting writers' intellectual property rights and permitting further distribution.

Sharing research data (data sharing) is a basic idea at the core of open science. This culture of openness and transparency transforms research by changing how academics work together, come up with new ideas, and add to the body of knowledge (Braunschweig et al., 2012).

Data sharing transcends the conventional research silos and reflects the spirit of openness and collaboration. It encourages the free exchange of research datasets, techniques, and analytical tools, enabling researchers, scholars, and PhD students to expand on one another's findings and discover previously uncharted territory in the field of knowledge creation. Transparency in research data sharing is essential to scientific rigour. It improves the repeatability of research results by enabling the confirmation and verification of findings. By allowing public access to datasets, scientists promote examination and cooperation, which increases the validity and trustworthiness of their findings. Research data is freely exchanged, which spurs creativity and quickens the rate of discovery—pivotal especially for PhD students. Shared datasets are useful tools that can be used to support multidisciplinary study, spark new questions, and spur innovations in a variety of sectors. Unrestricted access to data generates novel concepts that result in revolutionary breakthroughs in science and technology.

Ethical considerations are still crucial when promoting data sharing. It is essential to uphold the rights to privacy, secrecy, and intellectual property of both persons and researchers. Adhering to ethical norms and implementing appropriate data anonymisation procedures guarantee responsible and morally sound data-sharing practices.

Sharing data also presents difficulties. Coordinated efforts are needed to address concerns about security, privacy, and standardisation of data. Promoting an open data culture also requires encouraging researchers to share their findings, implementing appropriate attribution procedures, and acknowledging data-sharing initiatives.

Being aware of open science, open access, and data sharing (open data) is crucial for PhD students to understand how to work and disseminate in such open enviroments, taking advantage of those emerging possibilities.

Open science is based on the principles of open peer review and promotes transparency, accessibility, and collaboration. This novel technique of peer-review revolutionises the conventional academic evaluation procedure and improves accountability, transparency, and the calibre of scientific discourse.

Open peer-review increases the evaluation process's transparency by tearing down the barriers of anonymity. It promotes accountability by making reviewer identities and comments available alongside published articles. This enables readers to evaluate the legitimacy of the review process and authors to gain a deeper understanding of the criticism they receive.

Open peer review fosters scholarly discourse and constructive criticism. Openly disseminating reviewers' remarks and conversations promotes a more thorough and nuanced assessment of the literature and gives writers practical advice for enhancing their work.

The transparency of the peer-review procedure encourages thorough evaluation. Disclosure of reviewers' identities encourages them to deliver more thoughtful and in-depth evaluations, improving the calibre and scope of the review process. Scholarly collaboration is also encouraged by open peer review. It promotes conversations, debates, and teamwork among writers and reviewers, improving research results. It also gives reviewers recognition and credit for their contributions to the academic conversation.

Although open peer review has many advantages, there are drawbacks as well. Open peer review still needs to improve in the areas of creating standard operating procedures, guaranteeing reviewer confidentiality when needed, and striking a balance between openness and worries about bias or retaliation.

Preprints and open repositories are essential components of the open science framework that transform the impact, speed, and accessibility of scholarly communication (e.g. arxiv[25]). These platforms are a living example of openness; they allow researchers, particularly PhD students, to quickly publish their discoveries and add to the body of knowledge worldwide. Preprints provide a means for researchers to publish their findings quickly. Scholars can accelerate the interchange of ideas and discoveries by sharing their work with the scientific community and beyond through the uploading of manuscripts prior to formal peer review.

Open repositories make research outputs accessible. These platforms remove financial and geographic barriers by making academic publications, datasets, and other research materials publicly available. Important information is accessible to scholars, PhD students, learners, and the general public, promoting inclusive knowledge distribution.

Preprints and public repositories encourage cooperation and helpful criticism. Open sharing of research results allows PhD students to solicit comments, partnerships, and conversations from others, all of which can enhance the study process and produce higher-quality results. Preprint transparency increases the visibility and possible impact of research. Open access to shared findings facilitates

---

[25] https://arxiv.org.

early visibility and possible citations, which helps to acknowledge and disseminate research discoveries.

Open repositories and preprints have many benefits, but problems still exist. Due to concerns about quality control and the requirement for clear versioning of manuscripts, careful selection and unambiguous labelling of preprints are required.

In conclusion, adopting open scientific techniques provides a pathway towards increased collaboration, influence, and advancement in the academic environment for PhD students. It also educates them on transparent and ethical research. Early career engagement with open science principles can influence the direction of research and create a more diverse and influential academic community.

Teaching skills

PhD students are frequently hired as teaching assistants (TAs), which is certainly another source of funding to cover living expenses but also an important experience leading to teaching skills that are so rewarding and useful if planning a career in academia.

Teaching skills are introduced in this chapter for the natural relationship between excellent pedagogy and presentations that work. In order to convey knowledge effectively, teaching is a sort of presentation that requires clarity, involvement, and skilful communication. Doctorate candidates, who frequently work as TAs, have comparable difficulties while putting together a presentation, including organising the material, holding the attention of the audience, and explaining difficult concepts.

A TA's job is crucial to the academic environment since it acts as a link between students and teachers. As part of their academic career, PhD candidates frequently take on TA roles. TAs are responsible for assisting course instructors, facilitating conversations, evaluating papers, and assisting students with their learning. Learning good teaching techniques not only improves the learning environment for students but also helps to produce future researchers who can successfully manage both teaching and research responsibilities.

It is essential for Ph.D. candidates moving into TA roles to comprehend a variety of instructional approaches. This includes being acquainted with a range of instructional strategies such as conversations, lectures, active learning approaches, and the use of technology in the classroom. Workshops and training sessions can help PhD candidates understand how to adapt these techniques to a variety of learning styles and successfully convey difficult ideas.

Understanding pedagogical variety gives TAs a range of strategies to accommodate various learning preferences and classroom conditions. Various methodologies are available, including problem-based learning (PBL), active learning strategies, and flipped classroom tactics. TAs who are familiar with these frameworks are better able to determine which technique is best for a given topic, which promotes a dynamic and participatory learning environment.

A teaching strategy known as PBL puts students at the centre of their education and pushes them to solve hard, real-world issues. In PBL, students work on real, unstructured problems that reflect real-world or professional scenarios. As students work through these difficulties, this system promotes critical thinking, active participation, and teamwork.

The idea of self-directed learning is the foundation of PBL. As opposed to a conventional lecture-based method, PBL starts with the presentation of a problem or scenario that is frequently interdisciplinary in nature. Students are then required to use their prior knowledge and research skills to comprehend, evaluate, and come up with answers. Students gain an in-depth comprehension of the subject matter as a result of this process, which stimulates curiosity and the intrinsic drive to investigate.

The facilitator's job, which involves assisting students in their learning while granting them some degree of independence, is essential to PBL. As mentors or guides, facilitators provide assistance, ask insightful questions, and reroute conversations to keep students' attention on the major learning goals. PBL fosters collaborative teamwork since it requires students to participate in group discussions, exchange insights, debate viewpoints, and jointly design solutions. This cooperative setting fosters the development of communication skills, tolerance for differing opinions, and teamwork abilities—all of which are essential in future work environments.

However, PBL stresses applying information to solve complicated problems, which mirrors the difficulties people encounter in real-world situations. It goes beyond simple data memorising or recitation. Students who work through these issues gain not only subject-specific knowledge but also cross-disciplinary critical thinking, problem-solving, and decision-making abilities.

Finally, As an immersive teaching method, PBL pushes students beyond the mere assimilation of knowledge, encouraging critical thinking, cooperative problem-solving skills, and active engagement—all necessary for success in the complex and dynamic world of today.

A variety of approaches that go beyond conventional lecture-based instruction are included in active learning tactics. PBL is just an example of a method that motivates students to participate actively in their education. These tactics centre on providing students with interactive, participatory, and hands-on learning experiences in order to increase retention, stimulate critical thinking, and build engagement.

For instance, group conversations and cooperative activities are basic active learning techniques. Through peer discussion, debate, and information sharing, these workshops increase student involvement. These talks help students develop their communication and critical thinking abilities while also deepening their learning by creating an environment in which they may express their opinions, share their thoughts, and question one another's ideas.

Case studies and problem-solving activities are two commonly used active learning strategies. Through the use of complex problems or real-world circumstances, these exercises force students to apply their theoretical knowledge to real-world realities. Through the process of analysis, synthesis, and decision-making in these settings, students enhance their ability to solve problems, foster critical thinking, and close the gap between theory and practice. Additionally, students are immersed in practical experiences through interactive teaching methods, including role-playing, simulations, and experiential learning exercises. This gives them the chance to apply theoretical concepts in simulated or real-world scenarios. By allowing students to explore, experiment, and learn firsthand, these methods encourage active involvement and improve comprehension and retention of the material. Furthermore, the incorporation of technology is essential to active learning. Interactive multimedia materials, online learning environments, and educational technology tools provide opportunities for real-time feedback, individualised learning, and active participation. These resources support a variety of learning styles and increase student motivation and engagement by enabling self-paced learning, interactive exercises, and simulations. To sum up, active learning techniques operate as catalysts for changing the nature of education by enabling students to do more than just passively absorb knowledge.

A different teaching approach that emerged recently is the so-called flipped classroom approach, which is a cutting-edge method of instruction that reverses the conventional norm. In a flipped classroom, the conventional lecture-based method of content delivery is reversed: while in-person class time is devoted to active learning, group projects, and knowledge application, students interact with the course material outside of class, frequently through readings, recorded lectures, or online modules. The idea behind the flipped classroom is to use valuable in-person class time for problem-solving, deeper engagement, and interaction rather than just passively delivering knowledge. Before class, students are expected to access the learning materials on their own. This gives them the freedom to work at their own speed, review content, and get ready for more advanced discussions and activities when they meet in person. The basis for in-class activities is laid by pre-class learning resources like podcasts, videos, and online readings. Teachers use a range of interactive methods in the classroom to address students' questions, apply theoretical concepts in real-world situations, and reinforce comprehension. These methods include debates, group discussions, problem-solving exercises, case studies, and hands-on activities. This change in emphasis from learning that is instructor-centred to learning that is student-centred promotes peer interaction, critical thinking, and active involvement. Additionally, the flipped classroom model provides individualised learning experiences by allowing students to take charge of their education and accommodating a variety of learning styles. This method improves understanding, retention, and application of knowledge in the classroom by giving students the chance to engage in deeper research, ask questions, and work together to solve problems. The flipped classroom approach, which maximises class time for active learning and application by rearranging the learning environment and utilising technology, is essentially a paradigm change in education. Turning the conventional approach on its head gives students the authority to take an active role in their education, promoting critical thinking, deeper comprehension, and the development of critical skills that are vital for success in today's fast-paced, team-oriented world.

Using technology in instructional strategies is a modern strategy that works in the digital age. TAs can improve the learning process and meet the varied needs of contemporary learners by utilising interactive apps, online learning environments, multimedia resources, and instructional tools. TAs can design surveys, quizzes, and debates on Kahoot!,[26] a game-based learning platform that encourages participation through interaction and competitiveness. Akin to Kahoot!, Quizizz[27] provides gamified tests and quizzes for students to finish on their own, encouraging self-directed learning. With Nearpod,[28] TAs may design interactive classes that include real-time student engagement through multimedia content, polls, quizzes, and collaborative activities. Canvas[29] is a learning management system that provides features for a seamless online learning experience, including discussion boards, assignment submissions, material sharing, and grading tools. TAs can develop online courses with resources, activities, and evaluations using Moodle,[30] an open-source platform that promotes cooperation and communication between students and teachers. Google Classroom[31] offers an easy-to-use interface for sharing resources, arranging tasks, and promoting interaction between TAs and students.

As for multimedia resources, YouTube[32] offers a vast repository of educational videos, lectures, tutorials, and demonstrations across various subjects, supplementing course material and catering to diverse learning styles, while TED-Ed[33] offers engaging educational videos, lessons, and TED Talks curated for classroom use, sparking discussions and critical thinking among students. Finally, Khan Academy[34] provides a library of instructional videos, practice exercises, and

[26] https://kahoot.com/.
[27] https://quizizz.com/?lng=en.
[28] https://nearpod.com/.
[29] https://www.instructure.com/canvas.
[30] https://moodle.org/.
[31] https://edu.google.com/workspace-for-education/classroom/.
[32] https://www.youtube.com/.
[33] https://ed.ted.com/.
[34] https://www.khanacademy.org/.

personalised learning dashboards covering subjects from maths and science to arts and humanities.

There is also a plethora of instructional tools that might be useful for TAs, such as Padlet,[35] a collaborative digital board where TAs and students can share ideas, brainstorm, or compile resources collectively in a visually appealing format. Socrative[36] is a tool allowing TAs to create quizzes, assessments, and exit tickets to gauge student understanding and provide instant feedback, while Flipgrid[37] facilitates video-based discussions, enabling students to share thoughts, responses, or presentations, fostering a collaborative and inclusive learning environment.

Technology proficiency also enables TAs to adjust to online or hybrid learning environments, providing flexibility and inclusivity in the classroom. TAs who are proficient in adaptive learning software, such as Smart Sparrow,[38] can customise learning experiences for students by adjusting assessments and content according to each student's unique needs and performance. This promotes inclusivity and accommodates a variety of learning styles. Furthermore, Gradescope,[39] Turnitin,[40] and Microsoft Forms[41] can all be employed for the development and implementation of online tests, prompt feedback, and online evaluations that maintain academic integrity. There are also interactive tools that promote student engagement and creativity by enabling group idea-sharing, cooperative brainstorming, and problem-solving in a virtual environment, like Jamboard[42] and Miro.[43]

Another important aspect related to good teaching is effective communication. PhD students who want to become TAs should practice communicating concepts clearly and getting students involved in class discussions. Beyond only providing information, fostering empathy and approachability creates a welcoming atmosphere where students feel at ease asking for help. Mentoring abilities, such as the capacity to offer helpful criticism and assist students in their academic endeavours, are equally essential to the TA's job.

Concluding, it is crucial for PhD students who work as TAs to balance teaching and research responsibilities. Time management becomes critical for doctoral students navigating the multiple duties of research and education. Training sessions on time management techniques, work-life balance, and effective time allocation

---

[35] https://padlet.com/.
[36] https://www.socrative.com/.
[37] https://info.flip.com/en-us.html.
[38] https://www.smartsparrow.com/.
[39] https://www.gradescope.com/.
[40] https://www.turnitin.com/.
[41] https://forms.office.com/.
[42] http://jamboard.google.com/.
[43] http://miro.com/.

can help TAs balance their responsibilities. Fostering a supportive network to overcome the problems that come with this diverse profession can also be achieved by seeking collaboration and experience sharing among fellow PhD students working as TAs.

Summary

This chapter provided practical advice on how to organise, plan, and write a dissertation. It included advice on how to structure a dissertation, start and complete all chapters, deal with writing anxiety, and prepare for the final examination. It also included practical suggestions for supervisors on how to give appropriate feedback to students.

This chapter included all the elements related to presentation skills, both for the viva and for presenting to colleagues and faculties, giving talks at conferences and networking events. We learnt how to structure a presentation with the audience in mind, along with some practical advice for supervisors to help students plan for attending events and conferences.

This chapter offered guidance on disseminating a PhD student's results and collaborating with the supervisor and colleagues in writing academic papers. It included elements of writing style appropriate for a scholarly audience and public dissemination, as well as how to structure a paper and present the results of studies conducted during the PhD. It also included practical advice on how to get published in academic venues, which supplements the early career supervisor's advice. This chapter concluded with open science and the evolving landscape of research assessments. Considering that many PhDs work as TAs, it offered introductory guidance on lesson planning, curriculum design, and various teaching methods and strategies.

4

Assembling a PhD toolkit

This chapter serves as a guide to constructing a toolkit for doctoral research. It provides practical advice on different tools a student will need to carry out the work. This chapter includes tools for knowledge organisation, planning, data analysis, and writing, as well as reference management software and AI-based tools. Here is the list of tools—split into conventional and AI-based (see Table 4.1)—reviewed in this chapter, that can help students complete their PhD studies across various disciplines.

Bibliography

Reference management software is a programme that helps PhD students organise and manage the bibliographic references they use when writing academic papers, articles, and other scholarly works. The software typically allows PhD students to import references from various sources such as online databases and library catalogues, and then organise, annotate, and search the references. Some reference management software also includes tools for creating bibliographies and citations in a variety of formats such as MLA, APA, and Chicago.

Reference citation styles are standardised formats for how bibliographic references are presented in academic papers, articles, and other scholarly works. These styles dictate the order and punctuation of elements in a reference citation, as well as the use of italics, bolding, and other formatting. Some examples of commonly used citation styles include:

- MLA (Modern Language Association) style: typically used in the humanities such as literature and languages
- APA (American Psychological Association) style: often used in the social sciences, such as psychology and sociology, and other fields involving humans such as human-computer interaction or human-centred AI in the computer science field
- Chicago style: used in history and other disciplines that focus on the publication of books

The Doctorate Blueprint. Alessio Malizia, Oxford University Press. © Alessio Malizia (2025).
DOI: 10.1093/9780198927167.003.0005

Table 4.1 Conventional tools versus AI-based tools

Purpose	Conventional tools	AI-based tools
Bibliography	Endnote,[1] Mendeley,[2] Zotero[3]	scite.ai,[4] Grammarly[5]
Literature review	Litmaps.com,[6] Open Knowledge Maps[7]	Iris.ai,[8] Semantic Scholar,[9] Elicit.com,[10] ResearchRabbit.ai,[11] R Discovery[12]
Knowledge organisation	Cmap,[13] XMind,[14] MindMeister,[15] MindMup[16]	EdrawMind,[17] Taskade,[18] MyMap.ai[19]
Help writing and anti-plagiarism	Overleaf,[20] Google Docs,[21] Microsoft Word[22]	Grammarly,[23] Turnitin,[24] ProWritingAid,[25] QuillBot,[26] Jenni.ai,[27] PaperPal,[28] Scrivener,[29] Ulysses,[30] Hemingway Editor[31]
Time management	Pomodoro,[32] Thinkwell PhD toolkit,[33] Microsoft Project[34]	Trello,[35] Asana,[36] Forest,[37] Focus@Will,[38] RescueTime[39]
Data analysis	SPSS,[40] SAS,[41] R,[42] Stata,[43] Excel,[44] Google Sheets,[45] Python,[46] Jupyter Notebooks,[47] MATLAB,[48] Tableau,[49] Wolfram|Alpha,[50] NVivo,[51] Qualtrics[52]	RapidMiner,[53] MAXQDA,[54] RStudio,[55] IBM Watson Studio,[56] Microsoft Power BI[57]
Research skills	Vitae Research Development Framework[58]	Canva,[59] Perplexity,[60] ChatGPT,[61] Notion.so[62]

1 https://endnote.com/.
2 https://www.mendeley.com/.
3 https://www.zotero.org/.
4 http://scite.ai/.
5 https://www.grammarly.com/.
6 http://litmaps.com/.
7 https://openknowledgemaps.org/.
8 http://iris.ai/.
9 https://www.semanticscholar.org/.
10 http://elicit.com/.
11 http://researchrabbit.ai/.
12 https://play.google.com/store/apps/details?id=com.rdiscovery&hl=it&gl=US.
13 https://cmap.ihmc.us/.
14 https://xmind.app/.
15 https://www.mindmeister.com/.
16 https://www.mindmup.com/.
17 https://www.edrawmind.com/ai-mind-map.html.
18 https://www.taskade.com/.
19 http://mymap.ai/.
20 https://www.overleaf.com/.
21 https://www.google.com/docs/about/.
22 https://www.microsoft.com/it-it/microsoft-365/free-office-online-for-the-web.
23 https://www.grammarly.com/.
24 https://www.turnitin.com/.

Table 4.1 *Continued*

[25] https://prowritingaid.com/?gad_source=1&gclid=CjwKCAjwupGyBhBBEiwA0UcqaNIN9wKiQ
aya1XayY8Y8_4OCGksNX2ZPDbdZxEo4pp8B2Sh07JbCPhoCd0YQAvD_BwE.
[26] https://quillbot.com/.
[27] http://jenni.ai/.
[28] https://paperpal.com/.
[29] https://www.literatureandlatte.com/scrivener/overview.
[30] https://ulysses.app/.
[31] https://hemingwayapp.com/.
[32] https://www.toptal.com/project-managers/tomato-timer.
[33] https://www.ithinkwell.com.au/resources/PhDToolkit.
[34] https://www.microsoft.com/en-us/microsoft-365/project/project-management-software.
[35] https://trello.com/.
[36] https://asana.com/features/project-management.
[37] https://www.forestapp.cc/.
[38] https://www.focusatwill.com/.
[39] https://www.rescuetime.com/.
[40] https://www.ibm.com/it-it/products/spss-statistics.
[41] https://www.sas.com/it_it/home.html.
[42] https://www.r-project.org/.
[43] https://www.stata.com/.
[44] https://www.microsoft.com/it-it/microsoft-365/excel.
[45] https://www.google.com/sheets/about/.
[46] https://www.python.org/.
[47] https://jupyter.org/.
[48] https://it.mathworks.com/products/matlab.html.
[49] https://www.tableau.com/it-it.
[50] https://www.wolframalpha.com/.
[51] https://www.gmsl.it/nvivo/.
[52] https://www.qualtrics.com/it/.
[53] https://rapidminer.com/.
[54] https://www.maxqda.com/.
[55] https://posit.co/download/rstudio-desktop/.
[56] https://www.ibm.com/it-it/products/watson-studio.
[57] https://www.microsoft.com/it-it/power-platform/products/power-bi.
[58] https://www.vitae.ac.uk/researchers-professional-development/about-the-vitae-researcher-
development-framework.
[59] https://www.canva.com/it_it/.
[60] https://www.perplexity.ai/.
[61] https://chat.openai.com/.
[62] http://notion.so/.

- Harvard style: used in the fields of business, economics, and the natural
  sciences
- Turabian style: a version of Chicago style, specifically for student research
  papers

Each style has its own set of rules and guidelines, and it is important to use the
appropriate style for a given field or publication. Reference management software
like EndNote, Mendeley, and Zotero can help users format their references in
different styles.

EndNote,[1] produced by Clarivate Analytics, is a commercial reference management software package that is used to manage bibliographies and references when writing essays and articles. It can be used to search online bibliographic databases, organise references and images, and create bibliographies and citations in a variety of formats such as MLA, APA, and Chicago. It is available for Windows and Mac and also has a web-based version. It allows users to easily import references from various sources, such as online databases and library catalogues and then organise, annotate, and search the references. Additionally, it has features such as creating a library of references, formatting bibliographies in several styles, and collaborating with others.

Mendeley[2] is a free reference management software package that helps researchers and academics organise and manage the bibliographic references they use when writing academic papers, articles, and other scholarly works. It is available for Windows, Mac, and Linux, and it has a web-based and mobile version. It allows users to easily import references from various sources, such as online databases and library catalogues, and then organise, annotate, and search the references. Additionally, it has features such as creating a library of references, formatting bibliographies in different styles, and collaborating with others by sharing libraries, papers, and annotations. It also has a social network for researchers, which allows users to connect with other researchers in their field and discover new research papers.

When writing scholarly papers, articles, and other works, researchers and academics also utilise Zotero,[3] a free and open-source reference management tool, to organise and manage their bibliographic references. It offers a web-based and mobile version and is compatible with Windows, Mac, and Linux. It allows users to easily import references from various sources, such as online databases and library catalogues, and then organise, annotate, and search the references. The features it offers also include the ability to structure bibliographies in several forms, create a library of references, and collaborate with others by sharing papers, libraries, and notes. Additionally, it has an option that automatically finds and stores annotations and metadata from PDFs, web pages, and other documents. It also features a browser extension that makes it possible to add page citations to the library rapidly.

AI-powered tools for bibliography

Scite.ai,[4] an AI-powered platform, has emerged as a tool for helping PhD students discover, evaluate, and engage with research articles. By leveraging the Smart Citations feature, Scite.ai empowers students to uncover the nuances of research

---

[1] https://endnote.com/.
[2] https://www.mendeley.com/.
[3] https://www.zotero.org/.
[4] http://scite.ai/.

articles, assess their credibility, and contextualise their findings within the broader academic landscape, ultimately crafting their bibliography.

The Smart Citations feature provides PhD students with a comprehensive overview of how a research article has been cited by subsequent studies. This feature not only displays the frequency of citations but also classifies them as supporting, contrasting, or mentioning the original study. This approach to citation analysis enables students to quickly assess the impact and credibility of research papers, thereby streamlining the evaluation process and enhancing the overall quality of the bibliography and research.

Scite.ai's platform includes collaborative functionalities and supports PhD students in exploring research topics, identifying relevant articles, and engaging with their peers in a collaborative environment. The platform's features—such as saved searches and publication tracking—further help with assembling a bibliography.

Furthermore, Scite.ai's[5] Research Assistant is another key feature that sets the platform apart from other research tools. This tool leverages large language models (LLMs) to provide PhD students with real-time insights and suggestions, enhancing the research process. The Research Assistant is particularly useful for tasks such as generating search strategies and building reference lists that contribute to the bibliography.

A good video tutorial for Scite.ai[6] can be found on YouTube,[7] describing the main features of the tool and showing the use of the dashboard, which will be a handy tool for collecting references for the bibliography.

Grammarly[8] offers a suite of features that cater specifically to the needs of academic writers and, therefore, is useful for helping students write papers, which will be discussed later in this chapter. Nevertheless, one of the platform's standout capabilities is its automated citation generation, which seamlessly provides APA, MLA, and Chicago-style citations without requiring users to leave the web page they are working on. This streamlined approach to citation management can potentially save PhD students a lot of valuable time and reduce the risk of formatting errors, allowing them to focus on the content and quality of their work and ultimately helping in crafting a bibliography.

Literature review

Conducting a thorough literature review can often feel daunting but with the right tools, it can become a more engaging and rewarding experience. As Olesia Nikulina, PhD researcher, described in an X thread,[9] the following tools can help transform your literature review into an enjoyable process.

---

[5] http://scite.ai/.
[6] http://scite.ai/.
[7] https://www.youtube.com/watch?v=4tjfIJg40MY.
[8] https://www.grammarly.com/.
[9] https://tinyurl.com/ycxtd4ck.

Literature reviews are a fundamental component of academic and research endeavours, serving as the foundation for the exploration and advancement of knowledge. However, the process of conducting a comprehensive literature review can often be arduous and time-consuming. Litmaps.com,[10] an innovative online research platform, has emerged as a transformative tool that streamlines and enhances the literature review process. One of the key features of Litmaps.com is its ability to create interactive visual representations of literature maps. These maps provide PhD students with a comprehensive overview of the interconnected web of research articles that form the foundation of their topics of interest. By visualising the relationships between different publications, PhD students can gain valuable insights into the existing body of knowledge, identify gaps, and explore new research directions (Eppler, 2006). This visual approach not only enhances understanding but also stimulates creativity and critical thinking, similar to mind maps.

Litmaps.com offers a suite of tools that facilitate the organisation and collaborative aspects of research projects. PhD students can create literature maps for their specific projects, linking articles and references to build a cohesive narrative (Novak & Cañas, 2008). The platform's collaborative features enable PhD students to share their maps with team members, fostering real-time feedback and discussion. This collaborative environment promotes knowledge exchange and the development of a research community and is particularly useful if the PhD project is in a research laboratory.

Finally, Litmaps.com's database and search functionality empower students to find relevant literature more efficiently, saving time and effort in the research process. The platform's recommendation system, which suggests papers based on the student's interests and previous searches, further enhances the discovery of relevant sources and expands the student's knowledge base (Jannach et al., 2010). This personalised approach to literature searches and recommendations contributes to the overall quality and depth of literature reviews.

Open Knowledge Maps[11] significantly enhances the visibility of scientific knowledge for PhD students. It operates the world's largest visual search engine for research. This platform enables PhD students to create knowledge maps of research topics across all disciplines, providing an instant overview of a subject by visually displaying the main areas, relevant papers, and concepts associated with it. As mentioned by Nikulina, Open Knowledge Maps extensively visualises the research subject of interest by (1) displaying the primary research directions within it, (2) recognising all pertinent concepts and terminologies, and

---

[10] http://litmaps.com/.
[11] https://openknowledgemaps.org.

(3) grouping similar articles to facilitate comprehension. It is an essential tool for research exploration during the literature review phase.

At its core, Open Knowledge Maps operates on the principles of open science (more about it in this book), sharing source code, content, and data under open licenses. The platform serves as a building block of the open discovery infrastructure, collaborating extensively with other initiatives in the open science domain. The platform integrates optimised search, filter, and exploration functionalities tailored to different use cases and available metadata. PhD students can seamlessly transition between visual maps and synchronised list-based presentations, with the ability to view PDF documents and utilise the open annotation service Hypothes.is within the platform. It allows PhD students to view and submit annotations to PDF papers included in their research. It enhances the research experience by enabling students to interact with and annotate scholarly content directly within the platform.

Open Knowledge Maps offers a visually engaging and collaborative platform that empowers PhD students to explore, discover, and contribute to scientific knowledge. It embraces open science principles, fosters collaboration, and makes paper discovery useful during a literature review.

AI-powered tools for literature review

The process of conducting a literature study and discovering new information in the field of academic research has always been difficult and time-consuming. Iris.ai,[12] an AI-powered research platform founded by researchers from Stanford's Computation and Cognition Lab, leverages advanced language models and machine learning algorithms to automate and streamline the research process, empowering PhD students to navigate the vast expanse of scholarly literature.

At the core of Iris.ai's[13] features is the Researcher Workspace, a comprehensive platform equipped with a suite of tools designed to accelerate the literature discovery process. The platform includes five key modules—Explore, Analyze, Filter, Summarize, and Extract—each tailored to enhance different aspects of the research journey. By leveraging these modules, students can efficiently explore research topics, analyse document sets, filter relevant information, summarise key findings, and extract valuable data points.

One of the distinguishing features of Iris.ai[14] is its ability to understand the semantic relationships between concepts and ideas, enabling PhD students to

---

[12] http://iris.ai/.
[13] http://iris.ai/.
[14] http://iris.ai/.

uncover relevant publications beyond traditional keyword-based searches. By employing natural language processing techniques, Iris.ai[15] facilitates the discovery of thematic intersections and connections within the scholarly literature, allowing PhD students to identify novel research directions more easily. This semantic understanding enhances the depth and breadth of literature exploration, particularly in interdisciplinary fields where terminology and concepts may vary.

In addition to streamlining the literature discovery process, Iris.ai[16] offers tools to assess the reliability and quality of research papers. The platform's algorithms analyse various factors such as study design, sample size, funding sources, and potential conflicts of interest to provide users with a comprehensive reliability quotient for each publication.

Furthermore, Iris.ai's[17] collaborative features work quite well for knowledge-sharing among PhD students and research labs, enabling students and supervisors to engage in discussions, share insights, and collaborate on projects within the platform.

Semanticscholar.org[18] is an online platform that serves as a scholarly hub for researchers, students, and academics seeking to explore, discover, and engage with academic literature. With a vast repository of research papers, articles, and publications spanning various disciplines, Semanticscholar.org[19] facilitates research discovery and knowledge dissemination within the academic community.

Semanticscholar.org[20] offers users access to a diverse range of academic literature, enabling PhD students to delve into specific research topics, explore related works, and stay abreast of the latest developments in their fields of interest.

In addition to research discovery, Semanticscholar.org[21] serves as a valuable resource for academic writing support and guidance. The platform hosts a plethora of resources including articles, guides, and tools aimed at assisting researchers and students in honing their writing skills, crafting compelling research papers, and adhering to academic writing conventions. Whether it is navigating the intricacies of citation styles, structuring a coherent argument, or refining the clarity and precision of writing, Semanticscholar.org[22] offers a wealth of resources to support students at every stage of the writing process.

Through features such as comment sections, discussion forums, and collaborative tools, the platform encourages knowledge sharing, interdisciplinary dialogue, and collaborative research endeavours.

[15] http://iris.ai/.
[16] http://iris.ai/.
[17] http://iris.ai/.
[18] http://semanticscholar.org/.
[19] http://semanticscholar.org/.
[20] http://semanticscholar.org/.
[21] http://semanticscholar.org/.
[22] http://semanticscholar.org/.

One of the key strengths of Semanticscholar.org[23] is its commitment to quality assurance and reliability in academic literature. The platform employs rigorous vetting processes to ensure the credibility and integrity of the research papers and publications hosted on the site.

Elicit.com,[24] founded by researchers from Stanford's Computation and Cognition Lab, leverages the power of language models to automate and streamline the research process. By tapping into a comprehensive database of over 126 million academic papers across all disciplines, Elicit.com[25] empowers users to explore and synthesise relevant literature with speed and precision.

One key feature that sets Elicit.com apart is its ability to understand the semantic relationships between concepts and ideas rather than rely solely on keyword-based searches. This innovative approach allows users to uncover relevant publications that may not necessarily contain the exact search terms but are thematically aligned with the research question at hand. This capability is particularly valuable in interdisciplinary fields, where the terminology and conceptual frameworks may vary across different domains.

Moreover, Elicit.com's[26] AI-driven summarisation and data extraction functionalities enable PhD students to quickly identify the core insights and findings within a large corpus of literature. By breaking down complex research tasks into more manageable components, Elicit.com helps PhD students to synthesise information efficiently, identify key concepts, and extract relevant data.

One of the advantages of Elicit.com,[27] similar to Iris.ai, is its ability to assess the reliability and quality of the research papers it surfaces. The platform's algorithms analyse factors such as the study design, sample size, funding sources, and potential conflicts of interest, providing users with a 'reliability quotient' for each paper. This feature is particularly crucial in an era where the integrity and reproducibility of scientific research have come under increased scrutiny.

Furthermore, its collaborative features enable PhD students to share their work, receive feedback, and engage in discussions with peers, fostering a more collaborative and inclusive research ecosystem. This collaborative approach aligns with the principles of open science, which emphasise the importance of transparency, accessibility, and collective knowledge-building.

I would recommend to watching this video tutorial entitled 'How to use AI for research: Elicit.org for writing a literature review'[28] where some examples of

---

[23] http://semanticscholar.org/.
[24] http://elicit.com/.
[25] http://elicit.com/.
[26] http://elicit.com/.
[27] http://elicit.com/.
[28] https://elicit.com/?redirected=true.

how to conduct a literature review using Elicit.com[29] are shown, in particular the filtering feature. The video shows how PhD students can search for papers, read summaries of abstract to select sources, and filter keywords, publication date, and even study type, like meta-analysis or review.

ResearchRabbit.ai[30] is an AI-powered tool designed to help PhD students streamline the process of sourcing references for academic projects, essays, and literature reviews. Operating as a 'citation-based literature mapping tool', ResearchRabbit.ai[31] simplifies research, allowing PhD students to initiate their exploration with seed papers, from which the tool identifies additional relevant articles. While ResearchRabbit.ai[32] offers a good interface suitable for individuals with varying technical proficiency levels, it does have limitations, such as a database that may not be as comprehensive as platforms like PubMed or Google Scholar, potentially restricting the selection of available articles. Despite this, the tool facilitates the organisation of research articles and automatically generates bibliographies in various citation styles like APA, MLA, and Chicago, thereby simplifying the task of formatting references.

ResearchRabbit.ai's[33] visualisation feature provides students with a graphical representation of the academic network, illustrating relationships between scientific papers and co-authors. This visualisation aids in tracking specific topics or authors, enabling in-depth exploration of research efforts. Additionally, the tool offers concise article summaries and abstracts, assisting PhD students in quickly assessing the relevance of papers to their research.

While ResearchRabbit.ai[34] offers valuable features for literature review and academic research, it is important to note that access to the full range of tool features requires a subscription, which may pose a challenge for individuals with budget constraints. Furthermore, the tool lacks integration with other research tools and reference management systems, potentially inconveniencing users with established workflows. Despite these limitations, ResearchRabbit.ai[35] remains a good asset for PhD students seeking to streamline their literature review processes, enhance collaboration, and automate various research operations.

R Discovery[36] is an app designed to simplify the literature review process for PhD students and researchers. This AI-powered tool includes features for finding, organising, and reading academic research papers. Upon downloading the app, students can select the option 'Find papers for literature review' during the onboarding process, which enables a customised workflow tailored to their

[29] http://elicit.com/.
[30] http://researchrabbit.ai/.
[31] http://researchrabbit.ai/.
[32] http://researchrabbit.ai/.
[33] http://researchrabbit.ai/.
[34] http://researchrabbit.ai/.
[35] http://researchrabbit.ai/.
[36] https://play.google.com/store/apps/details?id=com.rdiscovery&hl=it&gl=US.

research area and topics of interest. R Discovery[37] then generates a personalised reading feed featuring relevant papers, allowing PhD students to bookmark and add papers to their literature review reading lists. The app also facilitates the export of selected papers in Microsoft Excel[38] and integrates with reference managers like Zotero[39] and Mendeley.[40]

One of the standout features of R Discovery[41] is its support for institutional access to paywalled articles, enabling students to read the full text of papers behind paywalls through their university's library affiliations. The app also provides access to over 40 million open-access journal articles, ensuring that PhD students have a vast pool of research papers at their disposal. Another key feature is R Discovery's[42] AI-generated summaries and highlights for around 6 million papers in its content bank. These allow PhD students to quickly assess a paper's relevance without having to read the full text.

By incorporating these tools into their literature review process, students can transform a potentially tedious task into an enjoyable and productive experience. I personally believe that the visual, organisational, collaborative, and interactive elements of these tools can help PhD students stay motivated, gain new insights, and ultimately produce a high-quality literature review.

Knowledge organisation

Students can effectively organise and synthesise their knowledge and research ideas using concept maps and mind maps. Links between research topics, theoretical frameworks, and research findings can be established through the utilisation of visual diagrams. While mind maps employ branches to connect similar concepts (e.g. tree-like structure), concept maps often use nodes or boxes to represent ideas and concepts (e.g. graph-like structure). These resources can be used by PhD candidates to organise their thoughts, develop ideas, and find research gaps. Visualising ideas in this manner facilitates a better understanding of how the various components of research fit together and enables the identification of areas requiring further investigation.

In this section, we will review some tools for concept and mind maps and introduce some principles on how to flesh out your knowledge and ideas using those tools.

---

[37] https://play.google.com/store/apps/details?id=com.rdiscovery&hl=it&gl=US.
[38] https://www.microsoft.com/it-it/microsoft-365/excel.
[39] https://www.zotero.org/.
[40] https://www.mendeley.com/.
[41] https://play.google.com/store/apps/details?id=com.rdiscovery&hl=it&gl=US.
[42] https://play.google.com/store/apps/details?id=com.rdiscovery&hl=it&gl=US.

CmapTools is a piece of software used to make concept maps. Concept maps are visual displays of knowledge that show the connections between concepts or ideas. The Florida Institute for Human and Machine Cognition (IHMC) created the programme, which may be downloaded[43] for free from their website. Using a number of different symbols, forms, and colours, PhD students may construct concept maps with CmapTools. Text, images, and hyperlinks can also be added to the maps. A number of tools are included in the software for managing and navigating sizeable maps and working together on map creation. CmapTools offers versatility, allowing the adaptation of concept maps to meet specific needs through the use of various templates, photo insertion, and incorporation of multimedia content.

Mind-mapping software allows PhD students to visually connect thoughts, ideas, and pieces of knowledge in a hierarchical and ordered way. XMind, Mind-Meister, and MindMup are three of the more commonly used mind-mapping software programmes.

- The flexible and effective mind-mapping software XMind[44] offers various functions such as brainstorming mode, Gantt charts, and presentation mode. XMind enables PhD students to customise colours, fonts, and themes to their preferences while also facilitating the addition of text, photos, and links to maps. The platform supports real-time collaboration and offers exporting capabilities in multiple formats, including PDF, Word, and Excel.
- The cloud-based mind-mapping tool MindMeister[45] provides simple choices for sharing and collaboration. With multimedia information like pictures, movies, and files, students may make interactive and interesting mind maps. With features like task management, comments, and presentation mode, MindMeister is suitable for team collaboration and project management. Additionally, MindMeister can be integrated with other programmes like Google Drive and Trello.
- Free mind-mapping software called MindMup[46] features an easy-to-use interface and is browser-based. PhD students may quickly develop and share mind maps and work in real-time collaboration with others. With MindMup, users can export their maps in a variety of formats, including PDF, PNG, and MindManager, as well as access features like task management, notes, and attachments. For students, especially those in a research lab, who require a straightforward mind-mapping application without the need for sophisticated functionality, MindMup is very easy to use.

---

[43] https://cmap.ihmc.us.
[44] https://xmind.app/.
[45] https://www.mindmeister.com/.
[46] https://www.mindmup.com/.

Some principles on how to take advantage of concept and mind-mapping software
A concept map demonstrates the connections between various ideas. An idea map is open-ended. There isn't a main concept in it. Each subject on the map must be a distinct, stand-alone concept (knowledge object). A line connecting one topic to another defines the nature of the link between the two topics. A concept map's text content cannot be effectively exported to, for instance, Microsoft Word. A concept map's text transferred to Word will be a useless collection of concepts. Instead, PhD students can organise and structure vast volumes of knowledge with a mind map. It aids in planning their work and memorisation by providing order. PhD candidates must arrange a lot of facts logically when writing a thesis, for instance. Mind maps, however, cannot assist them in comprehending the significance or meaning of a piece of information. Before adding the data to the mental map, it must be comprehended. PhD students might want to try creating an idea map if they are having problems understanding something. It happens frequently that PhD students start by trying to create a mind map but end up creating a concept map instead. This occurs when there are several lines without a hierarchy, and the concepts are not cleanly grouped into branches of primary themes and subtopics. Because concept maps don't give ideas a logical, linear sequence, the resulting graphic won't assist PhD students in formatting their work (e.g. paper or thesis chapter). Furthermore, concept maps produced by mind-mapping software are not tidy. A concept map made with mind-mapping software won't be understandable. PhD students must use concept mapping software to generate a concept map, just as you must use mind-mapping software to produce a mind map. The optimal approach to constructing a mind map for a dissertation involves inputting each research note directly into the mind-mapping application while conducting research and reviewing books or journal articles. Each note should start with a succinct title before the information is included in the text notes. Include the references for each note (e.g. a journal article, photo, or a scan of a book page). Attach the notes to the pertinent issues. PhD students might rephrase a few of their notes to create a more logical argument and save their mind map as a word processing document. To put it briefly:

- Directly enter notes as distinct ideas (individual knowledge objects) into floating topics.
- Include source documents (e.g. journal articles, web pages, recordings).
- Select the primary subjects for your task (sections headings).
- Join floating topics to the main topics.
- Verify the logical order of the ideas.
- Export to word processing and organise the text and argument (each paragraph should begin with an introductory sentence, a body, and a conclusion). The final sentence should refer to the first sentence of the following paragraph.

- Compose the introduction, body, and conclusion in a word processing software.

The best way to construct a concept map is to type each note made from books or journal articles into the idea mapping programme. Each note should start with a succinct title before the information is included in the text notes. Include the references for each note (e.g. a journal article, photo, or a scan of the book page). Summarising:

- Create individual knowledge objects for notes directly (separate ideas)
- Include source documents (e.g. journal articles, web pages, recordings)
- Select the primary subjects for the task (sections headings)
- Integrate floaters with main topics

AI-powered tools for knowledge organisation

The advent of AI-powered software tools for mind mapping can significantly enhance the cognitive processes of idea generation, organisation, and visualisation for PhD students. These tools, exemplified by EdrawMind AI and Taskade AI Mind Map Generator, leverage artificial intelligence algorithms to streamline the mind-mapping process, thereby helping PhD students manage knowledge organisation. By automating the creation of visually organised mind maps, these tools facilitate the structuring of thoughts and ideas, thereby enhancing cognitive clarity and facilitating more effective decision-making. Furthermore, the collaborative features of these tools, such as real-time co-editing and customisable templates, enable seamless group work and knowledge sharing, thereby fostering a more collective and participatory approach to idea generation and problem-solving for PhD students working in a lab, where they can share their knowledge organisation with the supervisory team.

EdrawMind[47] AI offers AI-powered mind-mapping capabilities that simplify the creation of mind maps. Advanced AI features include one-click mind map creation, AI Pre-Scene for various tasks, and Smart Annotation for adding notes efficiently. For instance, while creating a mind map, PhD students can right-click on a topic and select Create to generate subtopics in the form of branches, automatically filling in their mind maps. They can then modify and refine them by using other features to adjust the corresponding tree visually.

Taskade[48] AI Mind Map Generator Taskade AI Mind Map offers a valuable tool for PhD students embarking on the dissertation writing process. Students can

---

[47] https://www.edrawmind.com/ai-mind-map.html.
[48] https://www.taskade.com/.

utilise the Taskade[49] AI Mind Map in several strategic ways to maximise its bene-fits. Firstly, they can use the tool to brainstorm and organise their research ideas, concepts, and arguments in a visual and structured format. By creating a mind map that outlines the key components of their dissertation, students can gain clar-ity on the overall structure and flow of their work. Additionally, Taskade AI Mind Map can be employed to create detailed outlines for each chapter, section, or sub-section of the dissertation, helping students to break down complex topics into manageable segments. Furthermore, the tool's AI capabilities can assist in generat-ing connections between different ideas, identifying research gaps, and suggesting potential further explorations. Especially for mind maps, PhD students can use the integrated AI assistant in the form of a chatbot to generate, for instance, a mind map automatically by simply asking questions as they would when interacting with a standard LLM such as ChatGPT. For example, 'Can you help me structure and start writing a PhD dissertation on the use of AI in personal health management?' This will generate a table of contents with headings and subheadings, giving you a basic mind map to start structuring and writing a dissertation on such a topic.

MyMap.ai[50] is an AI-powered platform that offers a suite of tools to stream-line idea generation, organisation, and visualisation. One of its key features is the ChatMap, which leverages AI to facilitate mind mapping through a conversational interface. PhD students can simply describe their ideas or concepts to the AI assis-tant, and it will generate a visually appealing and structured mind map in real time. The ChatMap's ability to understand natural language and translate it into a coherent mind map saves students time and effort, making the ideation process more efficient.

In addition to the ChatMap, MyMap.ai[51] offers a powerful concept map tool that enables PhD students to create detailed, hierarchical representations of com-plex ideas and relationships. By breaking down concepts into their constituent parts and illustrating the connections between them, the concept map feature helps students gain a deeper understanding of the subject matter and identify potential areas for further exploration. The tool's AI-powered suggestions and auto-complete functionality make it easy to build comprehensive concept maps or mind maps, even for those PhD students with limited experience in mind mapping.

Another notable feature of MyMap.ai[52] is the NoteMap, which combines the benefits of mind mapping with the flexibility of note-taking. PhD students can create structured notes that are visually organised and easily navigable, making it simpler to capture and retain information. The NoteMap's ability to integrate

---

[49] https://www.taskade.com/.
[50] http://mymap.ai/.
[51] http://mymap.ai/.
[52] http://mymap.ai/.

with various file formats, such as PDFs and images, allows students to incorporate multimedia content into their notes.

In conclusion, AI-powered tools for knowledge organisation facilitate the way PhD students can approach idea generation, organisation, and collaboration. All of these AI-powered mind-mapping tools offer advanced features like one-click mind map creation, AI-driven smart features for brainstorming and organising ideas, and the ability to generate various types of mind maps quickly and effectively.

Writing

Writing is a crucial ability for PhD students to master in order to explain research ideas and findings successfully. There are a number of reasons why a PhD student might require writing tools:

- Productivity can be enhanced through the utilisation of writing tools, which offer features designed to aid in maintaining focus and achieving writing objectives. These features may include distraction-free interfaces, word count trackers, and project management functionalities.
- Quality improvement in writing can also be facilitated by the use of writing tools, which provide grammar and spelling checks, suggest more appropriate word choices, and help avoid common writing mistakes (e.g. Grammarly[53]).
- Writing tools can streamline collaboration, enabling easier interaction with advisors or research teams. These tools often support real-time document sharing, commenting, and revision tracking to ensure alignment and facilitate effective teamwork (e.g. Overleaf,[54] Google Docs,[55] Microsoft Word[56]).

Writing tools can generally make your PhD experience easier and more productive. They can assist you in avoiding wasting time and guarantee that your work is of the highest possible standard.

A thorough list of resources that writers and bloggers can utilise to enhance their writing skills is provided by Mary Jaksch (2023) in the article 'Top writing and blogging tools to enhance your writing skills'.[57] The article discusses a variety of tools, including online writing resources and writing apps.

Writing applications that assist authors with grammar, spelling, and style are the first group of tools addressed. Examples include Grammarly, Hemingway, and ProWritingAid. Tools for productivity and organisation, like Trello and Evernote,

---

[53] https://www.grammarly.com/.
[54] https://www.overleaf.com/.
[55] https://www.google.com/docs/about/.
[56] https://www.microsoft.com/it-it/microsoft-365/free-office-online-for-the-web.
[57] https://writetodone.com/top-writing-blogging-tools/.

fall under the second category and aid authors in managing their ideas and tasks. The Chicago Manual of Style and other online resources, which give authors access to a plethora of knowledge and direction on writing styles and procedures, make up the last group of tools addressed. In general, the post is a useful resource for PhD students who want to improve their writing abilities and efficiency.

It is crucial to expand on collaborative writing tools because they are a subject that is getting more and more attention in academic writing, considering that the majority of research produced nowadays is in teams. In order to enhance team-work and speed up the writing process, researchers and PhD students are looking for efficient solutions as the number of collaborative projects rises. PhD students may find collaborative writing to be a useful tool for improving their writing abilities, time management, and networking within their subject.

The three most popular platforms for group writing are Overleaf,[58] Google Docs,[59] and Microsoft Word.[60] While Overleaf[61] is a collaborative LaTeX editor largely used by scientists, researchers, and academics, Google Docs[62] is a cloud-based word processing programme that enables several users to edit a document concurrently, and so is Microsoft Word.[63] The ability for numerous users to collaborate on the same document at once is one of the key benefits of using collaborative writing platforms like Google Docs[64] and Overleaf.[65] As a result, team members, like PhD students working in a lab, can contribute to the project in real time with-out having to constantly communicate back and forth. This not only saves time but also lessens the chance of mistakes or misunderstandings.

These collaborative writing tools also allow users to track edits and updates made to a document, which is another benefit. In Google Docs,[66] for instance, PhD students and supervisors can post comments and suggestions for other lab members or between themselves to examine and view who made each edit and when it was made. Users of Overleaf[67] can quickly access earlier versions of the document and view a history of modifications and revisions made to the content.

Large and complex projects, such as writing a proposal or a journal paper, can be managed more easily with the help of collaborative writing tools. Authors can use Google Docs[68] to share particular files or folders with particular people or groups, and build folders and subfolders to organise their documents. Authors

---

[58] https://www.overleaf.com/.
[59] https://www.google.com/docs/about/.
[60] https://www.microsoft.com/it-it/microsoft-365/free-office-online-for-the-web.
[61] https://www.overleaf.com/.
[62] https://www.google.com/docs/about/.
[63] https://www.microsoft.com/it-it/microsoft-365/free-office-online-for-the-web.
[64] https://www.google.com/docs/about/.
[65] https://www.overleaf.com/.
[66] https://www.google.com/docs/about/.
[67] https://www.overleaf.com/.
[68] https://www.google.com/docs/about/.

using Overleaf [69] can also share their work with co-authors and collaborators, and organise their work into folders and subfolders.

Microsoft Word,[70] as part of the Office 365 suite, offers a set of features that can greatly assist PhD students in writing their dissertations and academic papers. One of the most valuable tools for PhD students is Word's citation and bibliography management system. By integrating with citation managers like Zotero and Mendeley, Word allows students to easily insert citations and generate bibliographies in various citation styles, such as APA, MLA, and Chicago. This feature helps with the referencing process, ensuring consistency and accuracy throughout the document.

Word's outlining and navigation tools are particularly useful for organising lengthy academic papers. The Outline view enables students to create and rearrange headings and subheadings, providing a clear structure for their dissertation or paper. The Navigation pane further enhances organisation by displaying all headings in a collapsible tree view, allowing students to quickly jump between sections and maintain a cohesive flow.

Additionally, Word's collaboration features, such as real-time co-authoring and commenting, facilitate teamwork between PhD students and their advisors or peers. These tools enable multiple users to simultaneously edit the document, provide feedback, and engage in discussions, ultimately improving the quality of the final work.

In conclusion, for PhD students and academics, especially if part of a lab or a research group, collaborative writing tools like Google Docs,[71] Microsoft Word,[72] and Overleaf[73] have become extremely useful. They enable real-time collaboration, keep track of updates, and simplify the management of extensive and intricate research outputs.

AI-powered tools for writing and anti-plagiarism

AI-powered writing tools have emerged as invaluable resources for PhD students navigating the challenges of dissertation writing and academic publication. These tools leverage artificial intelligence algorithms to provide personalised feedback, enhance writing quality, and streamline the research and writing process.

One of the primary benefits of AI-powered writing tools for PhD students is their ability to provide real-time feedback on writing quality, grammar, and

[69] https://www.overleaf.com/.
[70] https://www.microsoft.com/it-it/microsoft-365/free-office-online-for-the-web.
[71] https://www.google.com/docs/about/.
[72] https://www.microsoft.com/it-it/microsoft-365/free-office-online-for-the-web.
[73] https://www.overleaf.com/.

style. Tools like Grammarly[74] and ProWritingAid[75] use natural language processing algorithms to analyse text and identify areas for improvement such as sentence structure, word choice, and tone. This immediate feedback allows PhD students to refine their writing as they work, reducing the need for extensive revisions later in the process. Additionally, these tools often provide suggestions for alternative phrasing and explanations of grammatical rules, helping students develop a stronger command of academic writing conventions.

Grammarly[76] offers a comprehensive suite of features that cater specifically to the needs of academic writers. One of the standout capabilities of the platform is its automated citation generation, which seamlessly provides APA, MLA, and Chicago-style citations without requiring students to leave the web page they are working on. This approach to citation management saves PhD students time and reduces the risk of formatting errors, allowing them to focus on the content and quality of their work.

ProWritingAid[77] is an AI-powered writing tool that offers a wide range of features to assist PhD students in writing their dissertations and academic papers. With its AI-driven grammar checker, ProWritingAid[78] helps students perfect their work by identifying and correcting grammar errors, typos, and informal language, ensuring that their writing meets the high standards expected in academic writing. The tool's structure analysis, complex paragraph alerts, and advanced thesaurus tool are particularly beneficial for comprehensive reviews of lengthy academic papers, providing in-depth feedback on sentence structure, coherence, and academic rigour. Additionally, ProWritingAid's[79] analytical language goals and power verb suggestions help students strike a balance between clarity and academic proficiency in their writing so that their work is both academically rigorous and reader-friendly. Moreover, ProWritingAid's[80] citation tracking feature assists students in maintaining accuracy and consistency in their references, while the plagiarism detection tool helps ensure the originality of their work.

QuillBot.com[81] is an AI-powered writing tool that can greatly assist PhD students in crafting their dissertations and academic papers. One of QuillBot's most

---

[74] https://www.grammarly.com/.

[75] https://prowritingaid.com/?gad_source=1&gclid=CjwKCAjwupGyBhBBEiwA0UcqaNIN9wKi Qaya1XayY8Y8_4OCGksNX2ZPDbdZxEo4pp8B2Sh07JbCPhoCd0YQAvD_BwE.

[76] https://www.grammarly.com/.

[77] https://prowritingaid.com/?gad_source=1&gclid=CjwKCAjwupGyBhBBEiwA0UcqaNIN9wKi Qaya1XayY8Y8_4OCGksNX2ZPDbdZxEo4pp8B2Sh07JbCPhoCd0YQAvD_BwE.

[78] https://prowritingaid.com/?gad_source=1&gclid=CjwKCAjwupGyBhBBEiwA0UcqaNIN9wKi Qaya1XayY8Y8_4OCGksNX2ZPDbdZxEo4pp8B2Sh07JbCPhoCd0YQAvD_BwE.

[79] https://prowritingaid.com/?gad_source=1&gclid=CjwKCAjwupGyBhBBEiwA0UcqaNIN9wKi Qaya1XayY8Y8_4OCGksNX2ZPDbdZxEo4pp8B2Sh07JbCPhoCd0YQAvD_BwE.

[80] https://prowritingaid.com/?gad_source=1&gclid=CjwKCAjwupGyBhBBEiwA0UcqaNIN9wKi Qaya1XayY8Y8_4OCGksNX2ZPDbdZxEo4pp8B2Sh07JbCPhoCd0YQAvD_BwE.

[81] http://quillbot.com/.

valuable features is its paraphrasing tool, which allows students to rephrase sentences and paragraphs in various ways while maintaining the original meaning. This functionality is particularly useful for PhD students when summarising research findings or rephrasing passages from source materials to avoid plagiarism. In addition to paraphrasing, QuillBot[82] offers a grammar checker that identifies and corrects grammatical errors, typos, and improper word usage. This feature is essential for ensuring that dissertations and academic papers adhere to the high standards of academic writing expected at the doctoral level. QuillBot's[83] grammar checker can help PhD students polish their work and present their ideas clearly, concisely, and error-free.

Another key benefit of using QuillBot[84] is its citation generator, which automatically formats in-text citations and reference lists according to various citation styles, such as APA, MLA, and Chicago. Furthermore, QuillBot's[85] summarisation tool can help PhD students quickly extract the main ideas and key points from lengthy research articles and books. This feature is particularly useful during the literature review process, as it enables students to synthesise and incorporate relevant sources into their dissertations efficiently.

Jenni.ai[86] is an AI-powered writing assistant designed to help PhD students write their dissertations and academic papers. This app offers features that can enhance the writing process, improve productivity, and ensure the quality of scholarly work. One of Jenni.ai's[87] key functionalities is its AI-driven writing suggestions and corrections. Similarly to Grammarly,[88] this app uses natural language processing algorithms to analyse text, identify grammatical errors, suggest improvements in sentence structure, and provide feedback on writing style. This real-time assistance helps PhD students refine their writing, ensuring clarity, coherence, and adherence to academic conventions throughout their dissertations and papers.

Jenni.ai[89] also offers a plagiarism detection feature, which scans the text against a vast database of online sources to identify potential instances of plagiarism. By highlighting any unoriginal content, the app helps students maintain academic integrity and avoid unintentional plagiarism in their research work. Moreover, Jenni.ai[90] provides tools for organising and structuring academic writing. The app offers outlining features that allow PhD students to create detailed outlines for their dissertations, organise their ideas, and establish a logical flow of arguments.

---

[82] http://quillbot.com/.
[83] http://quillbot.com/.
[84] http://quillbot.com/.
[85] http://quillbot.com/.
[86] http://jenni.ai/.
[87] http://jenni.ai/.
[88] https://www.grammarly.com/.
[89] http://jenni.ai/.
[90] http://jenni.ai/.

Additionally, Jenni.ai[91] includes a citation management tool that assists students in generating accurate citations and bibliographies in various citation styles, such as APA, MLA, and Chicago. This feature simplifies the process of citing sources and ensures that students adhere to the required formatting guidelines in their academic papers.

Furthermore, Jenni.ai's[92] collaboration features enable PhD students to share their work with advisors and peers, facilitating feedback, revisions, and collaborative writing processes, especially if they are working in a research lab.

Like Jenni.ai,[93] PaperPal[94] provides tools for organising and structuring academic writing. One standout feature is that the app offers templates and guidelines for common academic document types, such as research papers, literature reviews, and dissertations. These templates help students establish a clear and coherent structure for their work, ensuring that their writing follows established academic conventions.

AI-powered writing tools also offer features that can enhance the overall structure and coherence of a dissertation or academic publication. Tools like Scrivener[95] and Ulysses[96] provide outlining and organisational features that help students break down their writing into manageable sections and chapters. These tools often incorporate AI-powered suggestions for structuring content, such as identifying potential subheadings or suggesting transitions between paragraphs. Additionally, some tools, like Hemingway Editor,[97] use AI to analyse the readability and clarity of a text, providing feedback on sentence length, passive voice usage, and overall flow.

While AI-powered writing tools offer numerous benefits, it is essential for PhD students to approach them with a critical eye. These tools should be used as a supplement to, rather than a replacement for, human feedback and editing. It is crucial for students to seek input from their advisors, peers, and writing centres to ensure that their work meets the specific requirements and expectations of their field. Additionally, students should be aware of the limitations of AI-powered tools and understand that they may not always provide accurate or contextually appropriate suggestions.

Avoiding plagiarism

Beyond citation formatting, Grammarly[98] also provides robust academic writing support through its plagiarism detection feature. By scanning the author's

---

[91] http://jenni.ai/.
[92] http://jenni.ai/.
[93] http://jenni.ai/.
[94] https://paperpal.com/.
[95] https://www.literatureandlatte.com/scrivener/overview.
[96] https://ulysses.app/.
[97] https://hemingwayapp.com/.
[98] https://www.grammarly.com/.

text against a vast database of online sources, Grammarly[99] can identify potential instances of plagiarism, enabling PhD students to maintain the integrity of their research and adhere to academic integrity standards. Additionally, the platform's proofreading and editing tools help students refine their writing, ensuring that their work is polished, coherent, and aligned with the conventions of academic discourse.

Turnitin[100] is a widely used software tool that helps PhD students ensure the originality of their dissertations and academic papers by detecting potential plagiarism. As a comprehensive writing support platform, Turnitin[101] offers a range of features that assist students in maintaining academic integrity throughout the writing process. One of Turnitin's primary functions is its plagiarism detection algorithm, which compares submitted documents against a vast database of online sources, including websites, journals, and previously submitted papers. This tool scans for verbatim matches, paraphrased content, and improper citation practices, providing students with a detailed report highlighting areas that may require further attention. By identifying potential plagiarism early on, Turnitin enables PhD students to make necessary revisions and properly attribute sources, ensuring that their work meets the high standards of academic honesty expected at the doctoral level.

In addition to plagiarism detection, Turnitin[102] offers a suite of writing support tools that help students improve the overall quality of their dissertations and academic papers. The platform's GradeMark feature allows advisors to provide detailed feedback on students' work, including comments, annotations, and rubric-based assessments. This feedback can help PhD students identify areas for improvement, refine their writing style, and strengthen their arguments, ultimately enhancing the quality and coherence of their final work.

Turnitin[103] also facilitates collaboration among PhD students, their peers, and their advisors through its PeerMark tool. This feature enables students to engage in peer review, exchanging feedback and suggestions with fellow researchers in a structured and organised manner. By participating in this collaborative process, students can gain valuable insights, identify potential weaknesses in their arguments, and learn from the experiences of their peers, all while maintaining the confidentiality of their work. Finally, Turnitin's integration with popular writing platforms, such as Microsoft Word[104] and Google Docs,[105] ensures seamless

---

[99] https://www.grammarly.com/.
[100] https://www.turnitin.com/login_page.asp?lang=it.
[101] https://www.turnitin.com/login_page.asp?lang=it.
[102] https://www.turnitin.com/login_page.asp?lang=it.
[103] https://www.turnitin.com/login_page.asp?lang=it.
[104] https://www.microsoft.com/it-it/microsoft-365/free-office-online-for-the-web.
[105] https://docs.google.com/document/create?hl=it.

integration into students' existing workflows. This compatibility allows PhD students to easily submit their work for plagiarism checking and feedback without disrupting their writing process.

Concluding, AI-based technologies can also improve research transparency and reproducibility. By automating data processing and analysis, these technologies can increase the reproducibility of studies and make it easier to share research approaches. This encourages transparency, enhances scientific integrity, and permits a more thorough examination and validation of research findings. However, it is crucial for PhD candidates to understand the ethical issues related to using AI techniques. PhD students should continue to be on the lookout for ways to protect the privacy of data, deal with biases, and defend intellectual property rights. PhD students can take advantage of the potential of generative AI tools while preserving their ethical integrity by integrating responsible AI practices into their study.

For instance, the ACM policy on authorship[106] states: 'Generative AI tools and technologies, such as ChatGPT, may not be listed as authors of an ACM published work. The use of generative AI tools and technologies to create content is permitted but must be fully disclosed in the work. For example, the authors could include the following statement in the acknowledgements section of the work: 'ChatGPT was utilised to generate sections of this work, including text, tables, graphs, code, data, citations, etc. If authors are uncertain about the need to disclose the use of a particular tool, err on the side of caution and include a disclosure in the acknowledgements section of the Work.'

Time management

For a PhD programme to be completed effectively, time management skills are crucial. To manage the demands of courses, research, writing, and other activities, meticulous preparation and organisation are necessary. Here are some suggestions for time management that I believe are useful for PhD students:

- Establish specific objectives: Clear objectives should be established for each level of the PhD programme, along with the identification of research objectives. Goals should be divided into manageable, smaller tasks, each assigned a realistic deadline.
- Task prioritisation: The to-do list should be sorted by priority and urgency. More time should be allocated to activities requiring greater attention or having shorter deadlines, with a focus on starting with the most critical tasks.

---

[106] https://docs.google.com/document/create?hl=it.

- Establish a schedule: A weekly or monthly calendar should be created to accommodate homework, other commitments, and research activities. Allocate specific time slots for each task, ensuring that breaks and rest periods are included.
- Utilise time-tracking apps, to-do lists, and productivity software as tools for productivity to assist in staying focused and effectively organising your time.
- Prioritise essential obligations and activities that align with research and academic goals, while politely declining non-essential commitments that may detract from focus. Utilise time effectively to accomplish tasks that contribute to academic objectives.

As a PhD student, employing these strategies can facilitate the development of efficient time management techniques, promote organisation, reduce stress, and improve academic progress.

Using time management strategies is crucial for being productive and reaching goals. The 60-30-10 rule and SMART goals are two widely used strategies. The 60-30-10 rule divides a daily routine into three halves as a time management strategy. The first section lasts for 60 minutes and is focused on a single task or undertaking. The second section, which lasts for 30 minutes, is utilised for breaks or to finish quick, easy activities. The final phase lasts 10 minutes, during which time the following 60-minute work session is planned. This method can help maintain motivation and productivity while mitigating the risk of burnout. I have found that using the 60-30-10 rule to organise time during a week works, too. Allocating 60% of the weekly schedule to high-priority tasks, such as submitting a paper to a conference, ensures that sufficient time is dedicated to critical activities with imminent deadlines. Around 30% of weekly time allocation can be allocated to mid-priority tasks, which encompass strategic activities lacking immediate deadlines, such as reviewing literature that may be relevant to the dissertation. Approximately 10% of weekly time allocation can be devoted to long-term objectives, characterised as low-priority tasks. These may include pursuits that can be found to be personally fulfilling but are not immediately urgent, such as exploring interdisciplinary ideas or writing a blog post to share personal perspectives.

On the other hand, SMART goals are specific, measurable, achievable, relevant, and time-bound. This strategy enables the definition of specific goals and objectives and the monitoring of progress towards achieving them. Creating measurable and well-defined goals allows for easy determination of when they've been achieved and facilitates strategy adjustments accordingly. For PhD candidates, SMART objectives are especially helpful since they can be used to divide large, difficult undertakings into smaller, more achievable tasks.

Here is an illustrative example of a SMART goal for a PhD candidate:

- **Specific:** Finish the chapter on literature for my dissertation, which is about 'The impact of climate change on computational sustainability'.
- **Measurable:** Study and synthesise at least 50 pertinent research publications and produce a 10,000-word chapter on the literature.
- **Achievable:** Devote at least four hours each day, five days a week, to reading, summarising, and writing in order to reach this objective.
- **Relevant:** The literature review chapter is a key component of a dissertation. It will give the research the backdrop and context it needs, and advance the larger academic debate on climate change and computationally intensive resources.
- **Time-bound:** Finish this goal in the next six months, which will give me adequate time to make changes and consider my advisor's advice before submitting the chapter for review.

Overall, this SMART goal gives the PhD applicant a specific and doable goal to strive towards and guarantees that progress can be monitored and assessed along the way.

Another good technique for focusing on objectives is the 'scary hour'. This expression, which has its roots in the hip-hop scene, is frequently used to describe the late-night hours when people are most productive. It is also frequently used to describe rappers who create their songs in the wee hours of the morning, when most people are asleep.

People frequently feel more motivated and focused during 'scary hour' since there are fewer interruptions and distractions. But it's crucial to remember that skipping sleep and staying up late might have a detrimental impact on your general health and productivity. It is advised that people place a high priority on getting enough sleep and developing a productive routine that suits their unique needs and way of life.

Personally, I believe that using a Pomodoro method within the 'scary hour' maximises productivity, especially if the scary hour is taken early in the morning.

The Pomodoro method, when used in conjunction with the idea of 'scary hour', can be a useful tool for maintaining concentration and productivity for a set amount of time. When using the Pomodoro method, one works for a certain amount of time—usually 25 minutes—followed by a brief rest. One might get a longer break after four 'Pomodoros' or work sessions.

During 'scary hour', tasks can be concentrated on for a fixed duration, followed by a brief break, and then the process can be repeated using the Pomodoro Technique. During periods when one may feel more focused and energised, utilising this approach can maximise productivity. However, prioritising sufficient sleep and not sacrificing well-being for job performance is essential. Establishing a reliable routine or adjusting schedules to accommodate more productive hours may be more sustainable and productive in the long term.

Gantt charts and PERT (programme evaluation and review technique) diagrams are two prominent project management tools for scheduling. Gantt charts make it simple to see what tasks are scheduled and when they need to be finished, since they give tasks and their timetables a visual representation. Another scheduling tool that emphasises the connections between jobs is the PERT diagram, which enables more accurate scheduling and risk assessments. Both methods can be helpful for time management.

Gantt charts can be made manually or with the aid of a number of software packages; more on those software packages later on in the chapter. They are useful for keeping track of task progress and making sure that everything is proceeding according to plan. A bar that represents each task and displays its start and end dates also displays its length. Dependencies can also be shown in Gantt charts, indicating which tasks must be finished before others can start. PhD students can quickly spot possible bottlenecks and assign resources as necessary by using a Gantt chart.

The critical path of a PhD study, or the sequence of tasks that must be finished on time to guarantee that the dissertation is completed on schedule, can be managed using PERT diagrams. As a result, scheduling and risk analysis (e.g. delays or supervisor's unavailability) are made more accurate, and students are better able to foresee potential delays and allocate time. By segmenting the work into smaller, more manageable tasks and then determining the dependencies between them, PERT diagrams are constructed. Students may then recognise the critical path and make sure it receives the proper priority.

Finally, PhD students can schedule and monitor the development of their work using Gantt charts and PERT diagrams, two efficient time management tools. Gantt charts are excellent for giving tasks and their deadlines a visual representation, making it simple to know which tasks are scheduled and when they need to be finished. On the other hand, PERT diagrams are used to determine a dissertation's critical path, which is necessary for efficient scheduling and risk analysis. Students may guarantee that their dissertation is finished on time and at the desired quality level of their work by employing these methods.

Several software tools, such as Microsoft Project,[107] Asana,[108] and Trello,[109] offer user-friendly interfaces for creating and managing Gantt and PERT diagrams. These tools often include additional features like task assignment, resource allocation, and progress tracking, making them valuable assets for PhD students in their time management efforts. By leveraging these software solutions, students can create and update their diagrams in real time, ensuring that their project plans remain relevant and adaptable to changing circumstances.

---

[107] https://www.microsoft.com/en-us/microsoft-365/project/project-management-software.
[108] https://asana.com/features/project-management.
[109] https://www.atlassian.com/it/software/trello.

Moreover, the use of Gantt and PERT diagrams can foster better communication and collaboration among PhD students, their advisors, and their peers, especially for students working in a research group or a lab. By sharing these visual representations, students can clearly communicate their plans, identify areas where support or guidance is needed, and engage in productive discussions about project progress and challenges.

Resources for PhD students are available on the Thinkwell PhD toolkit[110] website to aid them in overcoming the difficulties of their doctoral studies. That toolkit contains a range of resources, including worksheets, videos, and templates, on subjects including goal-setting and time management (e.g. Gantt and PERT), as well as academic writing and communication skills.

The toolkit's emphasis on developing students' self-awareness and thinking skills abilities is one of its essential components. This contains instruments for self-reflection on learning and development, as well as techniques for controlling feelings and sustaining motivation throughout the PhD process. The website also provides tools that are especially suited to the requirements of international students, who could experience additional difficulties as a result of linguistic and cultural differences. The useful and approachable resources offered can promote students' continuous personal and professional growth while also assisting them in navigating the challenging requirements of academic research and writing.

Finally, a good piece of advice can be found in the essay 'Project management for PhD students'[111] by Rachit Nigam (2019), which addresses the significance of project management expertise for PhD students. The author emphasises that a PhD is a substantial, intricate enterprise that requires meticulous planning and coordination to be successful. In order to stay on track and accomplish their objectives, PhD students need to learn project management techniques.

Nigam identifies four crucial project management competencies that PhD students need to possess. These include giving the project a defined scope and aim, breaking it down into manageable tasks, developing a plan and schedule, and regularly checking on progress. PhD students may make sure they are moving forward steadily towards their objectives and avoid becoming overwhelmed by the complexity of their project by employing these abilities.

The article also covers a number of methods and technologies that can be applied to help PhD students manage their projects. These include constructing a Gantt chart to visualise project timelines, utilising the Pomodoro method to better manage time, and using project management software like Asana[112] or Trello[113] to monitor tasks and deadlines. Nigam emphasises that these technologies should

---

[110] https://www.ithinkwell.com.au/resources/PhDToolkit.
[111] https://rachitnigam.com/post/project-management/.
[112] https://asana.com/features/project-management.
[113] https://www.atlassian.com/it/software/trello.

be used in conjunction with effective project management techniques, not as a replacement for them.

The article concludes with some advice on how PhD students can improve their project management abilities. These include enrolling in project management classes or seminars, looking for mentors or advisers with experience in the field, and proactively seeking feedback and critically evaluating one's own performance. PhD students can increase their chances of finishing their degree successfully and develop important skills that will help them in their future employment by honing these abilities.

AI-powered tools for time management

AI-powered tools for time management are AI-based tools that can assist PhD students in better time management, work prioritisation, and goal-keeping for their research. For example, Trello[114] is a project management solution that employs AI to help with task and project organisation and tracking. PhD students can make lists, move work through different phases, and set deadlines using the visual board structure it offers. Trello's AI features allow for automated task categorisation and intelligent task prioritisation suggestions.

The Forest[115] is a unique productivity app that employs gamification techniques and AI to help PhD students stay focused and manage their time effectively. Users can set a timer to work on specific tasks while the app grows a virtual tree. If the user exits the app before the timer ends, the tree dies. Forest's AI analyses user behaviour and provides insights into work patterns to optimise productivity and focus.

Another valuable AI-powered tool for time management is Focus@Will,[116] a productivity app that leverages neuroscience research to enhance concentration and productivity. By utilising AI algorithms to curate personalised music playlists that optimise focus and reduce distractions, Focus@Will[117] helps PhD students maintain a productive workflow during research and writing sessions. The app's adaptive technology adjusts music tempo and intensity based on individual preferences and cognitive patterns, creating an immersive work environment conducive to deep focus and sustained attention.

Additionally, RescueTime[118] is an AI-powered time-tracking tool that provides PhD students with valuable insights into their digital habits and time usage. RescueTime generates detailed reports on time allocation, productivity trends, and

---

[114] https://trello.com/.
[115] https://www.forestapp.cc/.
[116] https://www.focusatwill.com.
[117] https://www.focusatwill.com/.
[118] https://www.rescuetime.com.

potential distractions by monitoring computer usage, app activity, and website visits. This data-driven approach allows students to identify time-wasting activities, establish productive routines, and make informed decisions to optimise their time management strategies.

Data analysis

A PhD candidate necessitates proficient data analysis skills. Utilising statistical tools and procedures to extract crucial insights from data constitutes a fundamental aspect of research across various domains. Therefore, proficiency in data analysis methods and their application to address research problems is essential.

The ability to gather and clean data is a necessary data analysis skill. While cleaning data entails removing errors, inconsistencies, and missing numbers, collecting data entails locating sources of data and gathering them using the proper techniques. Prior to performing any statistical analyses, it is imperative to carry out this procedure to make sure the data are accurate and reliable.

The capacity to select the appropriate statistical technique for the current research question is another crucial data analysis skill for PhD students. The selection of the most appropriate statistical approach will rely on the type of data being used, the research topic, and the hypotheses being tested. This section will help students choose the best tool for their analysis; nevertheless, PhD students must be educated about various statistical methods and their underlying assumptions, which are normally offered as part of a basic training course on statistics and data analysis in a PhD programme or through participating in thematic summer schools.

Depending on the type and complexity of their research data, PhD students might use a variety of methods for data analysis. Here is a list of frequently used tools:

– Statistical software packages, such as SPSS,[119] SAS,[120] R,[121] and Stata,[122] are frequently used in various sectors, including the social sciences, health sciences, engineering, and business. These packages offer various statistical methods, including regression analysis, ANOVA, and factor analysis.

---

[119] https://www.ibm.com/it-it/products/spss-statistics.
[120] https://www.sas.com/it_it/home.html.
[121] https://www.r-project.org/.
[122] https://www.stata.com/.

- Data cleaning, processing, and visualisation are just a few of the basic data analysis operations that may be performed using the Excel[123] spreadsheet programme or Google Sheets.[124] Both include basic statistical analysis features.
- Python[125] is an effective programming language for statistical modelling, machine learning, and data analysis. Numerous libraries in Python, such as NumPy, Pandas, and Matplotlib, offer robust data analysis and visualisation capabilities, commonly employed through Jupyter Notebooks.[126]
- MATLAB[127] is a numerical computing environment frequently used to process data in engineering, physics, and other technical subjects. It includes built-in functions for statistical analysis, signal processing, and image analysis.
- PhD students can create interactive dashboards and visualisations with Tableau,[128] a data visualisation programme. This tool allows users to explore and analyse large datasets, which is especially helpful for producing visualisations for reports and presentations.
- Wolfram|Alpha[129] is a computational knowledge engine that provides factual answers and analyses in response to queries. It leverages a vast knowledge base and advanced data analysis algorithms to deliver results on a wide range of topics, including mathematics, statistics, geography, chemistry, and more.
- If qualitative data from focus groups, interviews, or open-ended survey questions are utilised in a study, tools such as NVivo[130] or Qualtrics[131] may be necessary. These software programmes offer tools for categorising and analysing qualitative data.

AI-powered tools for data analysis

PhD students can study massive datasets using AI-based tools that help them spot patterns or trends in the data. Some of the tools used for data analysis discussed in the previous section offer additional AI-powered features, so they also appear in this section.

One of the most widely used AI-powered tools for data analysis is SPSS[132] (statistical package for the social sciences). SPSS offers a user-friendly interface and

---

[123] https://www.microsoft.com/it-it/microsoft-365/excel.
[124] https://www.google.com/sheets/about/.
[125] https://www.python.org/.
[126] https://jupyter.org/.
[127] https://it.mathworks.com/products/matlab.html.
[128] https://www.tableau.com/it-it.
[129] https://www.wolframalpha.com/.
[130] https://www.gmsl.it/nvivo/.
[131] https://www.qualtrics.com/it/.
[132] https://www.ibm.com/it-it/products/spss-statistics.

a wide range of statistical analysis techniques, including regression, correlation, and factor analysis. The tool's AI-driven features include automated data cleaning, which identifies and corrects errors and inconsistencies in datasets. SPSS also offers automated variable selection, which helps students identify the most relevant variables for their analyses.

For PhD students working with large, complex datasets, tools like RStudio[133] and Python[134] offer powerful AI-powered capabilities. RStudio[135] is an integrated development environment (IDE) for the R programming language, which is widely used in data science and statistical computing. RStudio's[136] AI-driven features include automated code completion, which suggests relevant code snippets based on the user's input. The tool also offers automated testing, which helps students identify and fix errors in their code.

Python,[137] a versatile programming language, also provides a range of AI-powered tools for data analysis. Libraries such as NumPy, Pandas, and Scikit-learn offer advanced data manipulation and machine-learning capabilities. These libraries use AI algorithms to perform tasks such as data cleaning, feature engineering, and model building. Python's AI-powered tools are particularly useful for PhD students working with large, unstructured datasets, as they offer flexibility and scalability in handling complex data analysis tasks.

Another popular AI-powered tool among PhD students is NVivo,[138] a qualitative data analysis software. NVivo[139] uses AI algorithms to assist users in organising and analysing unstructured data, such as interview transcripts, focus group discussions, and social media posts. The tool's AI-driven features include automated coding, which identifies and categorises themes and patterns in text data. NVivo[140] also offers sentiment analysis, which determines the emotional tone of text data, providing valuable insights for studies in fields such as psychology, sociology, and marketing.

MAXQDA[141] is a powerful tool for PhD students analysing qualitative data. This software offers a comprehensive platform for managing, analysing, and visualising qualitative data, making it an indispensable resource for researchers in fields such as social sciences, education, and healthcare. MAXQDA's[142] AI-powered features include automated coding, which identifies and categorises themes and patterns in text data. The tool also offers sentiment analysis, which determines the

---

[133] https://posit.co/download/rstudio-desktop/.
[134] https://www.python.org/.
[135] https://posit.co/download/rstudio-desktop/.
[136] https://posit.co/download/rstudio-desktop/.
[137] https://www.python.org/.
[138] https://www.gmsl.it/nvivo/.
[139] https://www.gmsl.it/nvivo/.
[140] https://www.gmsl.it/nvivo/.
[141] https://www.maxqda.com/.
[142] https://www.maxqda.com/.

emotional tone of text data, providing valuable insights for studies in fields such as psychology, sociology, and affective computing, among others.

One of MAXQDA's[143] key strengths is its ability to handle large, complex datasets, including text, images, and audio/video files. The software's advanced data management capabilities enable students to organise and structure their data in a logical and coherent manner, facilitating the analysis process. MAXQDA's[144] AI-driven data visualisation tools, such as code maps and network diagrams, help students identify relationships and patterns in their data, providing a deeper understanding of their research findings.

One of the most prominent AI-powered tools for data analysis is Tableau,[145] a data visualisation platform. Tableau, through the use of AI algorithms, can assist PhD students in creating interactive dashboards and visualisations that help them understand complex datasets. The tool's AI-powered features include automated data preparation, intelligent recommendations for chart types based on data characteristics, and natural language processing capabilities that allow users to ask questions about their data and receive visual answers.

Another notable AI-powered tool is RapidMiner,[146] a comprehensive platform for data science and machine learning. RapidMiner's[147] AI-driven features include automated feature engineering, which selects and transforms the most relevant variables for predictive modelling. The tool also offers automated model selection, which tests various machine-learning algorithms and selects the best-performing model for a given task. RapidMiner's AI-powered text mining capabilities also enable PhD students to extract insights from unstructured text data, such as online questionnaires or social media posts.

IBM Watson Studio,[148] part of the IBM Cloud Pak for Data platform, is another powerful AI-powered tool for data analysis. Watson Studio provides PhD students with a suite of tools for data preparation, machine learning model building, and deployment. Its AI-driven features include automated data cleansing, which identifies and corrects errors and inconsistencies in datasets. Watson Studio[149] also offers automated model tuning, which optimises the hyperparameters of machine-learning models to improve their performance.

Depending on the amount of data collected by PhD students (e.g. analysing social networks or accessing big datasets), Microsoft Power BI,[150] a business

---

143 https://www.maxqda.com/.
144 https://www.maxqda.com/.
145 https://www.tableau.com/.
146 https://altair.com/altair-rapidminer.
147 https://altair.com/altair-rapidminer.
148 https://www.ibm.com/it-it/products/watson-studio.
149 https://www.ibm.com/it-it/products/watson-studio.
150 https://www.microsoft.com/it-it/power-platform/products/power-bi?market=it.

analytics service, also incorporates AI-powered capabilities. Power BI's[151] AI features include anomaly detection, which identifies unusual patterns or outliers in datasets. The tool also offers automated insights, which use natural language processing to generate explanations for key trends and insights in data visualisations.

Although there are many advantages to using AI-powered tools for data analysis, PhD students need to approach them critically. Instead of taking the place of human knowledge and discernment, these instruments need to enhance it. When working with sensitive or private data, students should be aware of the limitations of AI algorithms and make sure they are using the tools in an ethical and responsible manner. Students should also constantly confirm the dependability and correctness of their findings and be ready to defend their analysis techniques to peers and mentors.

Research skills

Any PhD programme must include teaching research skills because they are the basis for choosing a research topic and making original contributions to an area. PhD students have to master various skills, including information literacy, critical thinking, problem-solving, and data analysis.

The Vitae Research Development Framework[152] (RDF), a thorough and well-known tool for assisting PhD students and researchers in their personal and professional growth, is a good reference for planning how to develop such research skills properly.

The RDF comprises four core domains: engagement, influence, and impact; knowledge and intellectual abilities; personal effectiveness; research governance and organisation.

The core research competencies that PhD students, as future academics or researchers, need in order to conduct research effectively are the focus of the knowledge and intellectual abilities area. It encompasses skills like critical thinking, communication, and subject-specific knowledge. The focus of personal effectiveness is on the growth of character traits, including resilience, flexibility, and time management.

Research integrity, data management, and project management are just a few of the skills and expertise needed to ensure PhD students' efficient and ethical conduct. On the other hand, engagement, influence, and impact focus on the abilities required to interact with stakeholders, impact policy and practice, and disseminate research findings to a larger audience.

---

[151] https://powerbi.microsoft.com/it-it/.
[152] https://www.vitae.ac.uk/researchers-professional-development/about-the-vitae-researcher-development-framework.

Researchers at different career phases, from PhD students to senior academics, can use the RDF. It offers a flexible and adaptable framework for people to pinpoint their areas of strength and growth, create goals and targets, and plan their professional development. Universities and research organisations use the RDF to create and implement training and development programmes for researchers, as well as to assess and honour their accomplishments in these fields. Overall, the RDF[153] is useful for assisting PhD students in developing personally and professionally, allowing them to become more productive, involved, and impactful in their work.

AI-powered tools for assistance with research skills

Creating basic methodology or process flowcharts is highly beneficial for PhD students to improve their research skills and describe the research methods and processes used in their studies. These visual representations help students clearly communicate their research process, including data collection methods, analysis techniques, and the sequence of steps involved. Flowcharts provide a concise overview of the research methodology, making it easier for readers, such as thesis committee members or examiners, to understand and evaluate the study. By including well-designed flowcharts in their thesis, PhD students can enhance the clarity, organisation, and overall quality of their work, ultimately strengthening their research presentation and findings.

Canva[154] is a graphic design platform that can be a valuable tool for PhD students to create basic methodology or process flowcharts for their thesis. With its drag-and-drop interface and extensive library of templates and design elements, Canva[155] makes it easy for students to create visuals without prior design experience.

One of the key advantages of using Canva[156] for creating flowcharts is its availability of pre-designed templates specifically tailored for process diagrams and workflows. These templates serve as a starting point, allowing students to easily customise shapes, colours, and text to fit their specific needs. Canva's[157] library also includes a variety of icons and symbols commonly used in flowcharts, such as decision points, process steps, and connectors, simplifying the creation process.

---

[153] https://www.vitae.ac.uk/researchers-professional-development/about-the-vitae-researcher-development-framework.
[154] https://www.canva.com/it_it/.
[155] https://www.canva.com/it_it/.
[156] https://www.canva.com/it_it/.
[157] https://www.canva.com/it_it/.

Furthermore, Canva[158] offers a range of AI-powered features that enhance the design process for students. These features include Magic Write, Magic Presentation, Magic Design, Magic Edit, Magic Media, and Magic Translate, among others. Canva's[159] AI tools are grouped together within the software's 'Magic Studio' section, providing PhD students with easy access to these design-enhancing capabilities. One of these tools, Magic Write, is an AI-powered writing assistant that might help PhD students craft high-quality content for various purposes, such as academic papers. Magic Presentation allows students to create and edit presentations with effective data visualisations and a rich content library, starting with inspiration from professionally designed templates. The Magic Design feature generates multiple design templates based on a single idea, saving time and effort in choosing the right template. Magic Edit enables students to easily replace objects in photos with AI-generated replacements by brushing over the area and entering a prompt. Canva's AI image generator, Stable Diffusion, powers the Magic Media tool, which allows users to describe the image students might want to include in a presentation for a talk, for example, and the AI attempts to create it. Finally, Magic Translate is an AI-powered machine translation tool that instantly translates designs from 100 languages, but a Canva Pro license is required.

Perplexity,[160] as an AI assistant, can be a valuable tool for PhD students looking to enhance their research skills and streamline their academic work. Perplexity[161] offers a range of features and capabilities that can support students throughout the research process, from literature review to data analysis and thesis writing.

One of the principal ways Perplexity[162] can benefit PhD students is by providing personalised research assistance. The AI assistant can help students identify relevant sources, suggest research methodologies, and offer guidance on structuring their research projects. By leveraging Perplexity's[163] expertise and knowledge base, students can save time and effort in navigating the complexities of academic research.

One of the key features of Perplexity[164] is its ability to provide cited sources for every answer, ensuring the credibility and reliability of the information provided. The platform also offers a 'Copilot' feature, which is a guided AI search tool that enables students to explore topics in greater depth. Furthermore, Perplexity[165] allows users to ask follow-up questions, creating a thread of conversation that facilitates a deeper understanding of the topic.

---

[158] https://www.canva.com/it_it/.
[159] https://www.canva.com/it_it/.
[160] https://www.perplexity.ai.
[161] https://www.perplexity.ai/.
[162] https://www.perplexity.ai/.
[163] https://www.perplexity.ai/.
[164] https://www.perplexity.ai/.
[165] https://www.perplexity.ai/.

Perplexity[166] can also assist PhD students in conducting literature reviews. The AI assistant can help students search for and evaluate scholarly articles, identify key concepts and themes, and synthesise information from multiple sources. By leveraging Perplexity's[167] capabilities in information retrieval and analysis, students can ensure that their literature reviews are comprehensive, well-organised, and up to date.

Moreover, Perplexity[168] can support PhD students in data analysis and interpretation. The AI assistant can help students clean and preprocess data, perform statistical analyses, and visualise results. By providing guidance on data analysis techniques and tools, Perplexity enables students to derive meaningful insights from their research data and present findings effectively in their dissertations or academic papers.

In addition, Perplexity[169] can assist PhD students in improving their writing skills. The AI assistant can offer suggestions for structuring arguments, improving clarity and coherence, and adhering to academic writing conventions. By providing feedback on writing style, grammar, and citation practices, Perplexity[170] helps students enhance the quality of their written work and communicate their research effectively.

Furthermore, Perplexity's[171] project management features can help PhD students stay organised and on track with their research projects. The AI assistant can assist students in setting goals, creating timelines, and tracking progress. By providing reminders, deadlines, and task prioritisation, Perplexity[172] helps students manage their time effectively and meet project milestones in a timely manner.

Finally, Perplexity,[173] created by Anthropic, shares similarities with OpenAI's ChatGPT[174] in terms of being an LLM trained to engage in natural conversations and assist with a variety of tasks. Both Perplexity[175] and ChatGPT[176] leverage advanced natural language processing and machine learning techniques to understand and respond to user inputs. However, Perplexity[177] is a more recently developed AI assistant, and its capabilities may differ from ChatGPT's[178] in terms of knowledge cut-off date, training data, and specific features. While ChatGPT[179]

---

[166] https://www.perplexity.ai/.
[167] https://www.perplexity.ai/.
[168] https://www.perplexity.ai/.
[169] https://www.perplexity.ai/.
[170] https://www.perplexity.ai/.
[171] https://www.perplexity.ai/.
[172] https://www.perplexity.ai/.
[173] https://www.perplexity.ai/.
[174] https://chat.openai.com/.
[175] https://www.perplexity.ai/.
[176] https://chat.openai.com/.
[177] https://www.perplexity.ai/.
[178] https://chat.openai.com/.
[179] https://chat.openai.com/.

has gained widespread popularity and recognition, Perplexity[180] aims to provide a unique and valuable AI-powered research assistant tailored for PhD students and academics.

Prompt engineering for research skills

Prompt engineering is a crucial technique that PhD students can leverage when working with AI assistants like ChatGPT[181] and Perplexity[182] to enhance their research skills, literature review, academic writing, research methodology selection, and data analysis. By crafting well-designed prompts, students can guide these AI tools to provide more relevant and insightful responses, ultimately improving the efficiency and effectiveness of their academic work.

When using ChatGPT[183] and Perplexity[184] for research skills development, PhD students can employ prompts that prompt the AI assistants to provide guidance on research methodologies, data collection techniques, and literature review strategies. For example, prompts like 'Can you suggest research methods for a qualitative study on X topic?' or 'How can I improve my literature review process?' can elicit insights and recommendations from the AI assistants.

In the context of literature review, PhD students can utilise prompts to help them identify key sources, extract relevant information, and synthesise findings. Prompts such as 'Can you recommend seminal papers on X theory?' or 'How can I organise my literature review effectively?' can prompt ChatGPT[185] and Perplexity[186] to offer suggestions on literature search strategies, source evaluation criteria, and synthesis techniques.

For academic paper writing, prompt engineering can be used to seek advice on structuring arguments, improving writing style, and ensuring adherence to academic conventions. Students can craft prompts like 'How can I strengthen the argument in my paper?' or 'Can you provide tips for citing sources accurately?' to receive feedback and guidance from the AI assistants on writing quality and academic integrity.

When choosing research methods, PhD students can use prompts to seek recommendations on experimental design, data collection tools, and statistical analysis techniques. Prompts such as 'What are the pros and cons of using a mixed-methods approach?' or 'How can I ensure the validity of my research findings?'

---

[180] https://www.perplexity.ai/.
[181] https://chat.openai.com/.
[182] https://www.perplexity.ai/.
[183] https://chat.openai.com/.
[184] https://www.perplexity.ai/.
[185] https://chat.openai.com/.
[186] https://www.perplexity.ai/.

can prompt ChatGPT[187] and Perplexity[188] to provide insights into methodological considerations and best practices.

In the realm of data analysis, prompt engineering can help students receive guidance on data preprocessing, statistical tests, and visualisation techniques. By crafting prompts like 'How can I clean and prepare my data for analysis?' or 'What statistical methods are suitable for analysing survey data?' students can prompt the AI assistants to offer advice on data processing, analysis tools, and result interpretation.

Large language models such as ChatGPT[189] and Perplexity[190] rely on datasets that are the size of millions of books, web pages, and other language-related documents. Every word in these billions of sentences—or word fragments—becomes a token, and each token is then ranked in relation to every other token in the dataset based on how frequently it is used.

How do conversational abilities and complicated language come from the token-ranking approach used in language models? There is still research being done on the subject. However, learning how to manage this process doesn't require us to fully comprehend it. There are some guidelines though that might be useful for PhD students to prompt LLMs for enhancing their research skills.

Good advice is offered by an article by Michael Kranz, a post on the Notion.so[191] website. Notion is a productivity tool that combines note-taking, task management, and collaboration features into a single platform powered by AI features.

When utilising a generative AI model for your dissertation, remember that it differs from virtual assistants like Siri or Google Assistant. These models, trained on vast amounts of conversational data, grasp the intricacies of human interaction in text. To maximise its potential, engage with the model as you would with a person. By adopting a conversational approach, you can elicit more human-like and nuanced responses, enhancing the quality of your interactions and the insights gained from the model.

When making a request, include as much information as possible in the prompt, but keep it concise. Simpler language decreases the likelihood of misunderstanding.

Use positive phrasing. Avoid using negative phrasing when instructing an LLM, as it might focus on the action rather than the negation. For example, instead of saying, 'Do not include survey papers in the list', say, 'Only include papers containing experimental data in the list.'

---

[187] https://chat.openai.com/.
[188] https://www.perplexity.ai/.
[189] https://chat.openai.com/.
[190] https://www.perplexity.ai/.
[191] https://www.notion.com.

Attribute a clear identity to the model. For example, 'You are a PhD student collecting papers on topics X, in the area of subject Y' and then formulate a specific query. It will emphasise token patterns that are linked to actual PhD students inquiring about topics and subjects in an academic field.

One token at a time is how language models comprehend language. Concision is important since each token counts, but PhD students also can't rely on the model to correctly comprehend an ambiguous request, albeit it will give an output anyway. For example, a prompt such as: 'You are a PhD student; what data collection methods would you use?' will generate an answer list of some general-purpose methods like surveys, experiments, interviews, observation, focus groups, and so on. Instead, if the prompt is more specific, let's say: 'You are a PhD student in machine learning; what data collection methods would you use?' the output will be a list of specific methods such as web scraping, sensor data collection, crowdsourcing platforms, data logging, and so on.

Large volumes of data can be processed by LLMs. This implies that you might ask for a lot more writing prompts if you become proficient at creating them. Let's refine the prompt; for instance, 'You are a PhD student in machine learning; what data collection methods would you use if you had access to open datasets, including evidence to support your choices?' will generate methods and resources, including Kaggle, mentioning the evidence that many notable papers and contests have made use of Kaggle, web scraping suggesting BeautifulSoup or Scrapy as tools used when employing such methods or even the Amazon Mechanical Turk when choosing a crowdsourcing platform.

Enhance the prompt with clarifying sentences that address potential issues the model might encounter or choices it might need to make. For instance, if the research is focused on methods where there is a need to collect real-world data which is scarce or expensive to obtain (i.e. we need a form of simulated data), then we can add information to the prompt to steer away from unnecessary output: 'You are a PhD student in machine learning; what data collection methods would you use if you had access to open datasets, including evidence to support your choices but including only simulated data?' The answer generated by ChatGPT[192] becomes quite specific, even referencing a legitimate paper published on arXiv:

Method: Create simulated datasets based on known mathematical models and distributions. This approach is particularly useful in scenarios like physics-based simulations, financial modelling, and synthetic biology. Evidence: Simulated data is widely used in reinforcement learning, where environments are often simulated to train and test algorithms. The OpenAI Gym provides various simulated environments for training reinforcement learning agents.

(Brockman et al., 2016)

---

[192] https://chat.openai.com/.

When prompting LLMs for research, provide clear, specific, and diverse input-output examples to guide the model's understanding and generation. Include examples covering key concepts, methodologies, and findings from the specific field. Vary the examples' complexity, length, and style to expose the model to different formats. Regularly update and expand the example set to keep the model current as the research progresses. Continuing the example above, PhD students might prompt the chosen model by providing an example of data and format for simulated data in financial modelling:

Given the following example, how can I write a program to generate simulated data using Openai Gym for financial modelling? Example:

Date	Open	High	Low	Close	Volume
2023-01-01	150.0	152.0	148.0	151.0	10,00000
2023-01-02	151.0	153.0	149.0	152.0	1,100,000
2023-01-03	152.0	154.0	150.0	153.0	1,200,000
2023-01-04	153.0	155.0	151.0	154.0	1,300,000
2023-01-05	154.0	156.0	152.0	155.0	1,400,000
2023-01-06	155.0	157.0	153.0	156.0	1,500,000

Explanation
Date: the date of the trading day.
Open: the opening price of the stock for the day.
High: the highest price of the stock for the day.
Low: the lowest price of the stock for the day.
Close: the closing price of the stock for the day.
Volume: the number of shares traded during the day.

The output will be a detailed step-by-step guide with Python code and explanatory comments to generate this type of simulated data for financial modelling in an OpenAI Gym environment.

Overall, prompt engineering is a powerful technique that PhD students can use to maximise the utility of AI assistants like ChatGPT[193] and Perplexity,[194] among other LLMs in their academic work. By crafting well-designed prompts tailored to their specific research needs, students can access valuable insights, recommendations, and guidance to enhance their research skills, literature review process, academic writing, research methodology selection, and data analysis.

[193] https://chat.openai.com/.
[194] https://www.perplexity.ai/.

Supplementary readings

In assembling a PhD student's toolkit, it is advisable, in my opinion, to include three pertinent books as supplementary readings that can significantly aid throughout the PhD journey:

1. *Writing for computer science*, Zobel (2004), is a guide for computer scientists on how to effectively write research papers and technical documents. The advice is general enough to work for science and engineering most of the time.
2. *Research methods in human-computer interaction*, Lazar et al. (2017), covers various research methods used in human-computer interaction (HCI) that can be adopted whenever people are involved in research in science and engineering, including qualitative and quantitative methods, usability testing, and user-centred design.
3. *Experimentation in software engineering*, Wohlin et al. (2012), focuses on the experimental methods used in software engineering research to evaluate and improve software development processes.

Overall, these books aim to improve the research and communication skills of information systems and computing scientists but provide good general guidance on conducting effective research in various areas of science and engineering.

Writing for computer science by Justin Zobel (2004) is a comprehensive guide to writing technical documents and research papers in computer science. The advice in this book applies to different areas of science and engineering. The book provides practical advice and examples for computer scientists and PhD students to improve their writing skills and communicate their research effectively.

The book covers various topics, including how to write abstracts, introductions, and conclusions, as well as how to structure and organise technical documents. Zobel also provides guidance on how to write clear and concise sentences, how to use diagrams and graphs effectively, and how to avoid common writing pitfalls.

Jonathan Lazar, Jinjuan Heidi Feng, and Harry Hochheiser's book, *Research methods in human-computer interaction* (2017), provides a thorough manual for conducting research in HCI and information systems. Still, it helps in general when involving people and stakeholders in research in science and engineering. The book covers a wide range of research techniques, such as surveys, case studies, experimental designs, and both qualitative and quantitative techniques.

The authors offer helpful guidance on how to properly plan and carry out studies, as well as how to select the best research method for a particular research subject. The book also discusses data analysis methods, how to report research findings, and ethical issues in research.

Experimentation in software engineering by Claes Wohlin et al. (2012) provides a thorough manual for carrying out experiments in software engineering. The book covers the planning, design, implementation, and analysis of experimental research, which is general enough to be useful for PhD students in science and engineering.

The authors offer helpful advice on formulating pertinent research questions, selecting and measuring variables, planning experiments, and evaluating and presenting the findings. The book contains examples of experiments done in several branches of software engineering, including requirements engineering, testing, and maintenance, as well as ethical aspects of experimental research. For PhD students in the field of science and engineering who aim to enhance the calibre and efficacy of their work through experimental methods, *Experimentation in software engineering* is a useful resource.

Summary

This chapter offered practical advice on different tools a student might need to carry out the work. It included conventional and AI-powered tools for knowledge organisation, planning, data analysis, writing, and presentation, as well as reference management software.

This chapter offers PhD students a comprehensive guide to leveraging AI-powered tools and software to enhance various aspects of their research and writing process.

It begins by exploring how AI-powered tools can revolutionise the literature review stage. By automating the gathering, cataloguing, and synthesis of relevant sources, these technologies can save students considerable time and effort, allowing them to delve deeper into the theoretical and empirical foundations of their work. For instance, the chapter lists specific platforms, such as Zotero and Mendeley, that seamlessly integrate bibliographic management with knowledge organisation.

Building on this foundation, the chapter then delves into the role of AI in structuring and managing the wealth of information accumulated during the research process. Intelligent note-taking apps, mind-mapping software, and knowledge graph visualisers are discussed as means of transforming disparate ideas into a cohesive, accessible framework. This section emphasises the importance of maintaining intellectual control over one's research, even as the volume of data grows.

Recognising that effective writing is central to the doctoral journey, the chapter dedicates a substantial portion to AI-powered writing assistants and anti-plagiarism tools. These technologies can help students improve their academic prose, ensure proper citation, and maintain originality—all while boosting productivity and confidence.

Time management emerges as another critical area where AI can make a significant impact. Furthermore, the chapter explores AI-powered task managers, calendar integrations, and productivity trackers that can help PhD students prioritise their workload, minimise distractions, and stay on top of deadlines.

The final sections of this chapter delve into the role of AI in data analysis and research skill development. Sophisticated visualisation tools, statistical modelling software, and prompt engineering techniques are presented as means of extracting deeper insights from PhD students' research, as well as enhancing core competencies such as literature searching and critical thinking.

5

The future

What's next

Congratulations! You have successfully finished the arduous process of earning your PhD. You have accomplished a big academic milestone after years of challenging research, restless nights, and numerous experiments. However, you may find yourself asking, 'What's next?' as you stand on the verge of fresh opportunities.

This chapter, titled 'What's next', seeks to answer that question and give PhD students useful information about the many options available to them moving forward. While earning a PhD opens up a wealth of prospects, it is imperative to approach this new stage with caution and foresight. This chapter will discuss important decisions and offer helpful guidance on the most important elements of the post-PhD experience.

There are four sections in this chapter, each of which focuses on a different facet of future efforts. PhD students will have a thorough idea of the several paths their career can follow after exploring these parts:

- Developing a career plan: As doctors begin this new phase of their professional lives, it is crucial to create a strategic career plan. Utilising the information in this area to assist in determining interests, strengths, and objectives can establish a clear career trajectory. By contemplating various career options both within and outside academia, individuals can better equip themselves to make decisions aligned with their goals.
- Writing a grant proposal: Acquiring financial backing for research endeavours is a pivotal step in one's academic journey. Developing proficiency in grant-writing techniques proves immensely advantageous, irrespective of one's path within or beyond academia. This chapter delves into the intricacies of composing a robust grant proposal, offering guidance on optimising the prospects of securing funding for research projects.
- Establishing and running a research lab can be a rewarding career for those who have an interest in academia. This chapter will explain the duties and difficulties involved in operating a research lab, such as hiring staff, obtaining financing, and creating a welcoming and effective research atmosphere. Individuals will acquire insights into effective leadership strategies and gain valuable guidance on achieving success in such roles.

The Doctorate Blueprint. Alessio Malizia, Oxford University Press. © Alessio Malizia (2025).
DOI: 10.1093/9780198927167.003.0006

- Most PhDs do not become professors, while many doctoral graduates may dream of doing so. The truth is that there aren't many open opportunities in academia. The chapter will delve into the data, providing a comprehensive understanding of the wide range of career opportunities available beyond academia. PhDs will be introduced to various career paths that align with their unique skill sets, including industrial research, consultancy, science communication, and entrepreneurship.

Developing a career plan

Even if the path ahead may appear unclear, taking the time to consider personal interests, strengths, and long-term objectives will offer clarity and guidance. Creating a professional strategy requires self-reflection, research, and proactive decision-making. By taking the actions mentioned in this section, PhDs will be prepared to make decisions that are in accordance with their goals and increase their chances of success.

All those actions break down into multiple steps:

- *Self-assessment:* It involves considering one's abilities, talents, and passions, as well as determining values, interests, and long-term professional objectives. Understanding personal motivators and identifying preferred work environments can guide individuals towards aligning job decisions with their personal and professional goals.
- *Exploring career options:* It involves investigating and considering various job opportunities available to PhD graduates. Beyond academia, potential career paths may include roles in business, government, non-profits, or consultancy. Attending employment fairs, seminars, and networking events provides opportunities to connect with professionals from diverse fields and gain insights into different career trajectories.
- *Skill development:* Identify any knowledge or skill gaps relevant to the intended job path for *skill development*. Seek opportunities for professional development, such as workshops, courses, or certifications, to enhance existing talents or acquire new ones. Consider gaining real-world experience through internships or group projects to broaden the skill set.
- *Networking:* Attending relevant conferences, seminars, and workshops within the field allows for connections with scholars, experts, and potential employers. Participation in online forums and discussions related to the areas of interest is also recommended. Networking efforts may lead to mentorship, collaborative opportunities, and potential job prospects. Platforms such as LinkedIn and ResearchGate offer valuable opportunities for networking.

- *Seek mentorship*: Find mentors who can offer advice and assistance during job transitions. Find seasoned experts who can provide guidance, share their experiences, and offer insightful commentary. A mentor may help overcome obstacles, make wise choices, and introduce you to beneficial possibilities and resources.
- *Set goals*: Create a clear, practical career strategy with both short- and long-term objectives. The objectives should be broken down into manageable steps with deadlines. Review and re-evaluate the objectives frequently to make sure they continue to reflect personal, changing interests and aspirations.
- *Stay flexible*: Be adaptable to change and welcome new opportunities. The professional landscape is ever-changing; therefore, it's critical to continue to be adaptable and flexible. Be prepared to explore new avenues, take reasonable risks, and seize chances that fit with personal interests and objectives.

A straightforward and useful method for career planning is the '5-Minute Career Action Plan'[1] from jobs.ac.uk. The plan's goal is to assist individuals in developing a clear understanding of their career objectives and in making concrete efforts to realise them. It has four important steps.

Individuals are first urged to consider their present circumstances and evaluate their talents, shortcomings, values, and aspirations. They are able to determine their professional priorities and what they really want to accomplish through self-reflection.

Setting definite, defined goals is the second phase. These objectives should be difficult yet attainable and in line with the person's personal and professional ambitions. A balanced and upward-moving career trajectory is ensured by setting short-, medium-, and long-term goals, according to the plan.

The third step focuses on determining the processes and resources required to meet the objectives. This includes determining any skill gaps and figuring out how to fill them, looking for applicable training or educational options, interacting with professional networks or mentors who can offer advice and assistance, and more.

Taking action is required for the 5-Minute Action Career Plan's final phase. People are urged to divide their goals into doable projects and assign due dates to each activity. To keep the plan relevant and flexible to changing conditions, it is also strongly recommended that it be reviewed and updated on a regular basis.

Overall, the 5-Minute Career Action Plan offers a succinct framework for considering doctoral grads' career aspirations, creating goals, finding the resources they need, and taking action to achieve their goals. By devoting a brief amount of time to this plan, doctoral grads can obtain clarity and direction for their careers, empowering them to make decisions with knowledge and move closer to their intended professional results.

---

[1] https://www.jobs.ac.uk/media/pdf/careers/resources/the-5-minute-career-action-plan.pdf

Writing a grant proposal

The ability to write grant submissions is crucial for PhD holders looking for funding for their research initiatives. Listed below are some tips for doctoral grads about writing strong grant applications, along with some literature that can give some more detailed advice (Browning, 2022; Karsh et al., 2019):

Understand the grant guidelines: Learn about the exact rules and specifications of the grant you are applying for. Pay particular attention to the scope of the grant, eligibility criteria, deadlines, and any specific formatting or submission instructions. Check the grant's goals and objectives to see if the proposed research fits within the scope of the grant. Make sure the project is within the grant's thematic or focal area scope. Examine the eligibility criteria, such as academic standards, citizenship limitations, or links with specific institutions. Before submitting the application, make sure all the requirements are met. Take note of the application deadline, as well as any relevant deadlines for submitting letters of intent or draft proposals. To properly organise progress and ensure timely submission, create a timeline. Observe any specific formatting instructions given, such as font size, page restrictions, or necessary parts. Pay close attention to the suggested file type (such as Word or PDF) and any additional paperwork that needs to be submitted.

Clearly define the proposed research project: It's crucial to give a precise and succinct description of the research project in a grant request that emphasises the following features: clearly explain the project's precise goals and research questions. What do you hope to learn or accomplish with your research? Make sure the aims are clear and in line with the grant's objectives.

Stress the significance and applicability of the study. Describe the significance of the project and how it fills a knowledge gap or advances the field. Draw attention to any possible effects or conclusions from the research. Describe the anticipated results or deliverables of the research, usually under the heading 'Expected outcomes'. This could include written works, visual displays, data collections, software applications, or other specific results. Regarding the goals your project wants to accomplish throughout the grant's allotted period, be precise and reasonable. Clearly state the knowledge gap or shortfall that the research endeavour seeks to address. Describe how the effort expands on prior research, advances the topic, or fills a gap in the literature. Cite relevant literature to back up the claims.

Conduct a thorough background research: Conduct a thorough analysis of the scholarly literature that is pertinent to the research topic. Determine the key theories, ideas, and prior studies that form the basis of the proposed endeavour. To make sure that the proposed review reflects the most recent understanding of the subject, pay attention to recent publications. Locate any questions that remain unresolved or places where more investigation is required. This emphasises the relevance and originality of the proposed idea. Emphasise how the proposed research closes these gaps or contributes to the body of knowledge by doing so. Cite prior research to highlight the significance and applicability of the proposed study. Describe how the proposed project advances prior research or offers fresh perspectives that may have an influence on the field. This makes the case for sponsoring the proposed study or research stronger. Use reliable academic sources to back up every assertion. Following the citation format described in the grant rules, include in-text citations and construct a list of references at the end of the submission.

Identify key references: Select references that support the suggested project and are closely related to the research goals. Search for papers, books, or articles that relate to the particular subject, approach, or theoretical underpinnings of the investigation. Prioritise recent articles to show knowledge about the most recent developments in the field of study. While foundational works are vital, more recent research demonstrates familiarity with the field's most recent developments and the present level of knowledge. Give peer-reviewed publications preference. Peer-reviewed articles go through a thorough review process by subject-matter specialists, assuring their dependability and calibre. This increases the legitimacy of the proposed research. References from respected journals, well-known academic institutions, eminent scholars, or other authoritative sources should be used. These references support the claim and give it credence. Give peer-reviewed publications preference. Peer-reviewed articles are subjected to meticulous assessment by professionals; choose sources that fairly represent a variety of viewpoints, approaches, and findings on the proposed topic. This indicates expertise in various theoretical perspectives and research methodologies in the field. Follow the citation rules for the grant, such as APA, MLA, or Chicago style, when referencing sources in the grant proposal. Make sure that page numbers, journal abbreviations, author names, and publication titles are formatted correctly and consistently.

Include a well-structured methodology: Explain in detail the research strategy or approach that will be used for the investigation. Indicate the type of approach used, whether experimental, observational, qualitative, quantitative, or a combination of methodologies. Explain why this design is suitable for achieving the research goals. Specify the approaches and processes that will be used to acquire data. Surveys, interviews, experiments, observations, and dataset analysis could all fall under this category. Describe how these techniques will be used to collect the data needed to answer the research questions. Describe the statistical or analytical approaches that will be employed to analyse the data. Explain how these methods can be used to examine the research variables and achieve the research goals. Mention any specialist software or tools planned for use, if applicable. Recognise any obstacles or restrictions in the process. This could involve problems with data collection, participant access, equipment, or any other elements that could influence how the research will be carried out. Describe ways to overcome or lessen these difficulties. Make sure the suggested methodology can be implemented within the suggested timescale and with the available resources. Take into account the grant's parameters and whether or not the suggested study plan can actually be carried out. Reviewers can learn if the idea is well planned and feasible by seeing evidence of its viability.

Address potential limitations: Spend some time critically analysing the project to find any potential flaws or limits that could affect the results or application. Constraints relating to sample size, data accessibility, methodology, time, or any other elements that can affect the reliability or generalisability of the project may fall under this category. Outline the exact tactics or procedures that will be taken to mitigate or lessen the impact of any potential restrictions once those are recognised. For instance, if problems with sample size are expected, workarounds to this constraint might be explained by using the right statistical tools, working with other academics, or using other data-gathering strategies, that is using a SWAT analysis method. SWOT analysis is a strategic planning tool used to evaluate the strengths, weaknesses, opportunities, and threats of an individual, organisation, or project. A relevant book on the topic of SWOT analysis is *The handbook of improving performance in the workplace* (Leigh, 2009). It is crucial to underline the importance and possible contributions of the research, even if it has some limitations. Describe how, despite the noted restrictions, the research will nevertheless offer insightful new information or close a significant gap in the literature. Emphasise the distinctive features of the strategy or the fresh viewpoint offered to the study issue. If severe restrictions are evident and may be difficult to overcome within the project, then seek collaboration or input from subject-matter experts. The practicality and robustness of the research can be improved by working with colleagues or using specialised resources, successfully addressing any potential limits. Demonstrate the ability to modify and adapt the research strategy in the event that project constraints change unexpectedly. This exhibits a capacity for critical thought and the flexibility to adapt when necessary, ensuring the effectiveness of the research as a whole.

Develop a realistic budget: Break down the budget into distinct categories and then list the projected costs for each area. This could involve paying salaries to employees, purchasing equipment, paying for supplies and materials for research, compensating participants, paying for publications, or any other relevant costs. Explain why each expense in the budget is required for the successful completion of the research project. Give a detailed rationale for every expense in the budget. Justifications may be based on supplier bids, market prices, past performance, or any other pertinent facts. This aids reviewers in determining whether the budget is realistic. Make sure the budget complies with the requirements and limitations established by the sponsoring organisation. Be mindful of any specific spending restrictions, permitted expenses, or other monetary specifications mentioned in the grant rules. This demonstrates that the proposal has given significant thought to the financing constraints. If necessary, take into account indirect costs or over-head expenditures related to the proposed research. Administrative fees, facility usage fees, and other expenses that support the infrastructure required for the research infrastructure are examples of indirect costs. Verify that indirect expenditures are permitted by the grant and include them properly. Indicate specifically how the desired award fits into the overall funding strategy if funding is requested from several different sources. Show how the requested grant will be used to pay for specific project costs and how it works in conjunction with or in addition to other funding sources.

Seek feedback and review: Grant proposals should be shared with trustworthy colleagues, mentors, or experts in the field for their comments and suggestions. They can offer insightful commentary, point out potential areas for development, and make recommendations to improve your plan. Take into account those who have written grants before or have reviewed grants of a similar nature. Hold sessions or discussions in collaboration with other researchers or members of your research team. The conversations that result from this collaborative process may help enhance the proposal. Consult researchers or people who have previously been successful in obtaining grants, if at all possible. They can offer direction on how to complete the grant application process, share their personal experiences and views, and offer advice particular to the funding organisation or grant programme. Their knowledge can help improve the proposal and make it more competitive. Think about hiring a qualified editor or proofreader to check the proposal for style, grammar, coherence, and clarity. An impartial eye can spot mistakes, make the proposal easier to read, and make sure the reviewers understand the point of view offered by the proposal. Take into account the advice and criticism received. Adapt and improve the main idea in light of feedback from reviewers. To boost the proposal even further, address any issues or gaps found during the assessment process shared with the reviewers.

Edit and proofread: Make sure ideas are presented in a clear and coherent manner by reading over the proposal. For better readability, look for any assertions that are unclear or contradictory. Make sure the proposal follows a logical, easy-to-follow flow that leads the reviewers through the research's aims, methodologies, and predicted results. Verify the text for any grammatical mistakes, paying particular attention to subject-verb agreement, verb tense consistency, and punctuation usage. Be sure to word correctly and be consistent throughout the proposal. Use grammar and spell-checking software and language tools such as Grammarly,[2] but be careful because they might not capture all mistakes. Examine the proposal's formatting and structure. Verify that the sections, headings, and subheadings are all designated consistently. Make sure that the proposal complies with the necessary formatting standards, including font size, line spacing, and margins.

---

[2] https://app.grammarly.com/

A well-designed and organised proposal presents itself as more professional and is simpler for reviewers to read. Consider your proposal's total length and make an effort to be brief. Eliminate any extraneous repetition, flowery language, or material that is redundant. To effectively communicate your thoughts, choose language that is clear and simple. Keep in mind that reviewers frequently have a limited amount of time to assess several proposals, so a concise and straightforward proposal might be effective. Ensure that references and in-text citations are correct, regular, and adhere to the prescribed citation style. Verify again that the reference list has every source mentioned in the text and vice versa. Check that the author names, publication dates, titles, and other information are accurate. Citations that are incorrect or lacking in key information can cast doubt on your proposal's veracity. Think about asking someone else to look through the proposal, such as a mentor or colleague. They can offer a different viewpoint and point out any mistakes or places that might need clarification. The opinions of a third party might be quite helpful in enhancing the proposal's overall quality.

Finally, remember that writing grant proposals is a skill that gets better with practice. Be ready to modify and improve your proposals based on feedback and experience.

Running a research lab

It is critical for a recent PhD graduate to comprehend the motives and potential advantages of managing a research lab. Despite the difficulties and obligations that come with the job, running a lab offers a chance for scientific research and discovery. The course of the work can be determined, a group of outstanding individuals can be assembled, and intriguing scientific problems can be addressed by leading a research group. Assuming the responsibilities of a lab leader enables the promotion of collaboration, the training of new scientists, and contributions to the broader scientific community.

The article 'What your PI forgot to tell you: Why you actually might want a job running a research lab'[3] by Amy and Mark (2017) discusses the benefits for a PhD student thinking about a career in research lab leadership and offers a good piece of advice.

The capacity to create and foster a future vision is an essential component of managing a research lab. Important experience and information are gained during training as a PhD graduate. Transitioning into a lab leader position allows for the utilisation of this knowledge to advance research goals. The opportunity arises to work on cutting-edge research, explore fresh concepts, and broaden understanding of the chosen subject. Leading a lab offers the potential for a lasting influence

---

[3] https://www.molbiolcell.org/doi/full/10.1091/mbc.e17-02-0091

on the field of study, providing the freedom to determine research approaches, secure funding, and establish partnerships.

Running a research lab offers the chance for both professional and personal development. Abilities in project management, communication, and mentoring can be enhanced by assuming a leadership role. A comprehensive understanding of constructing interdisciplinary collaborations, navigating scientific funding complexities, and guiding lab members' career paths can be acquired. The position also facilitates interaction with others in the field through conferences, partnerships, and scientific communities, contributing to scientific career advancement.

Finally, managing a research lab provides access to a variety of professional routes. Although academics is a frequent career route, a lab leader's abilities and expertise can be applied to a variety of scientific and non-scientific positions. In business, government, policy-making, or other fields, the capacity to manage projects, coordinate teams, and communicate effectively can be extremely significant. It is crucial to take into account the potential for personal fulfilment and impact in various industries, as well as the long-term career chances that come with managing a research lab.

Creating a future vision

The importance of recognising and fostering a future vision as one progresses into the next stage of their career cannot be overstated. The trajectory of their research will be influenced by their ability to develop and nurture a vision, which will also impact their lab colleagues, collaborators, and the broader scientific community.

It is advisable to dedicate time to reflect on their research passions and identify the overarching objectives and issues that drive them. Clearly articulating the vision involves outlining the specific issue or topic they intend to explore, considering the potential significance of their findings, and detailing the innovative strategies they plan to employ. This serves as a foundational step in developing a comprehensive and compelling vision.

Consider dedicating a significant amount of time to reviewing literature pertaining to the area of interest. Analyse the contributions of esteemed scholars who have made substantial impacts in the field. By examining their methodologies, approaches, and research trajectories, insights into the formation of their perspectives can be gained. Substantiate the assertions and demonstrate expertise by referencing relevant scholarly sources.

Engage actively in discussions and collaborative endeavours with fellow scientists to refine and broaden the scientific outlook. Seek input from mentors, colleagues, and subject-matter experts to challenge the assumptions and expand the perspectives. Participate in interdisciplinary exchanges and idea-sharing platforms at conferences, seminars, and workshops to refine your unique worldview.

While assuming greater leadership responsibilities, embrace the role of advising and guiding the lab team. Encourage them to develop their own scientific visions aligned with the lab's overarching objectives. Foster a supportive work environment conducive to collaboration and mutual support. By granting team members the autonomy to pursue their research passions, a shared vision that fosters innovation can be cultivated.

A vision is dynamic and changes in response to new knowledge, technological development, and societal demands. Adopt a growth mentality, and remain open to changing and honing the vision as new information comes to light. Keep up with new developments in design, science, and technology that may influence the specific outlook. Make sure the objectives are in line with the developing academic, societal, and industrial landscape by periodically re-evaluating and redefining them.

The one-minute manager

In my opinion, one of the most useful resources for managing a research lab is Kenneth and Spencer's (1983) timeless classic, *The one minute manager*. This book provides insightful advice and doable leadership strategies. The writers of this book offer a straightforward yet effective management philosophy that emphasises attaining the best results through effective communication, goal-setting, and positive reinforcement.

Three essential strategies—one-minute goals, one-minute praising, and one-minute reprimands—comprise the book's main theme:

- Setting specific, succinct, and measurable goals is important, according to the authors. They stress the importance of clearly outlining goals so that people may comprehend them and successfully monitor their progress. Leaders can provide their team members with clarity and focus by setting up one-minute goals, which will allow them to direct their efforts towards attaining specified objectives.
- One-minute praise: Recognising and rewarding positive behaviour is an essential component of effective leadership. The writers promote prompt, sincere appreciation when team members reach or exceed objectives. Expressing gratitude for the person's efforts, highlighting their unique successes, and making the praise meaningful and genuine are all part of the one-minute praise process. This method promotes a happy workplace, increases motivation, and improves general job satisfaction.
- One-minute reprimands: Addressing errors and departures from standards is a crucial component of efficient management. The writers offer the idea of 'one-minute reprimands', which entails addressing problems quickly and

constructively while still showing respect for the person. Leaders may help team members improve their performance and evolve by providing feedback in a respectful yet forceful manner. One-minute reprimands promote a culture of support and learning while helping to retain accountability.

For instance, an example of setting a one-minute goal can be detailed as follows:

Goal: The objective is to 'create a cutting-edge digital platform for the preservation and analysis of historical texts from the Renaissance'.

Description: The objective is to create a digital platform that will make it possible for academics and researchers to access, study, and preserve historical documents from the Renaissance. To improve the study and comprehension of these extremely valuable cultural artefacts, the platform should combine cutting-edge digitisation techniques, natural language processing algorithms, and interactive visualisation capabilities.

Measurement: Reaching the following benchmarks within the allotted timeframe will show whether this goal was successful. Examples of benchmarks:

1. Curate and digitise a sizable collection of texts from the Renaissance, including written and printed works.
2. Create an intuitive user interface that enables academics to browse, search, and annotate the digitised texts.
3. Apply natural language processing techniques to the digitised resources to extract metadata, pinpoint significant themes, and run text analysis.
4. To enable users to explore and visualise patterns, linkages, and historical settings within the texts, develop interactive visualisation tools.
5. Create partnerships with academics and institutions to grow the collection and support the growth of the user base for the digital platform.

A one-minute praise and reprimand in the same scenario would sound like this:

Praise: 'Carrie, I wanted to take a moment to thank you for your outstanding contribution to our research project in the digital humanities. You did a fantastic job digitising the Renaissance manuscripts with thorough data curation and attention to detail.'

Description: Carrie has distinguished herself as a research lab member by putting extraordinary effort and skill into digitising and organising Renaissance manuscripts. His rigorous approach and focus on detail made a huge difference in the collection's quality and use.

Impact: Thanks to Carrie's efforts, important historical texts are being preserved and are now available to researchers worldwide. Her outstanding effort has raised the bar for future digitisation initiatives and improved the reputation of the research centre. Additionally, her attention to detail made proper metadata extraction possible and increased the digital platform's general usability.

Reprimand: 'Sarah, I want to talk about a problem with the accuracy of the metadata extraction during the digitisation process. To sustain our digital platform's dependability and usefulness, we must retain high-quality data.'

Description: Sarah has been in charge of the research lab's metadata extraction efforts throughout the digitisation process. On occasion, the accuracy of the retrieved metadata has been shown to be inaccurate, which negatively affects the dependability and usability of the digital platform.

Discussion: The lab manager talks about the particular problem of extracting metadata and underlines the significance of accuracy in preserving the integrity of the digital collection. The rebuke is delivered in a constructive and encouraging way, emphasising the gravity of the situation without undermining Sarah's worth or potential.

Most PhDs are not professors

The myth that most PhD holders choose to work as professors in higher education is probably wrong. Actually, many employment options and professional routes are open to PhD holders outside of traditional academia. In my personal experience, the majority of the PhD graduates I supervised work outside the academia; there is a caveat, although it depends on the country: in some, there is little market for PhD graduates outside the academia, but things are definitely changing over time with national programmes, for instance, offering tax discounts for private research centres or industries hiring a candidate with PhD titles.

PhD programmes traditionally educate students for careers in research and academia, but in practice, there are many more PhD graduates than there are open faculty seats in universities. As a result, a lot of PhD holders discover that they are interested in pursuing various job pathways in business, government, non-profit, and industry.

There are many reasons why PhD holders choose careers outside of academia. The need for a work-life balance, the accessibility of more varied employment options outside of academia, the requirement for financial security, or a preference for applied research in industrial settings could all be among these causes.

PhD graduates frequently engage in extensive academic research for several years, which can necessitate major personal sacrifices. Many people yearn for a more balanced living as they near the end of their PhD studies. A better work-life balance is promised by non-academic occupations, which also have more stable work schedules and less pressure to publish and obtain funding. PhD graduates who choose careers outside academia might devote more time and effort to hobbies, family obligations, and personal interests, improving their overall well-being.

Although academia offers a stimulating intellectual environment, there may be fewer career alternatives available there than in non-academic fields, which offer a much wider range of chances. PhD graduates have a special set of abilities that are highly sought after outside of academia, including critical thinking, problem-solving, and research expertise. Advanced degree holders are actively sought after by sectors such as technology, consulting, finance, healthcare, and government organisations to promote innovation and make evidence-based decisions. PhD graduates can investigate various fields, put their expertise to use in real-world situations, and make a significant contribution to solving problems.

Furthermore, students who pursue a PhD typically have to make financial compromises because they devote a lot of time and money to their studies. Many graduates find themselves looking for better financial stability after years of earning meagre stipends or relying on subsidies. Non-academic occupations may provide more enticing pay scales, perks, and long-term growth opportunities than academic ones. PhD graduates benefit from this stability and freedom to follow their passions and personal ambitions without being constrained by financial obligations.

While academic institutions are known for their emphasis on basic research, some PhD graduates have a strong preference for applied research in business or industry contexts. These people are drawn to the chance to cooperate on projects with real-world applications, work closely with business experts, and see the immediate results of their studies. PhD holders can bridge the gap between theory and practice by conducting applied research in non-academic settings. This enables them to turn their expertise into real-world results that enhance technology, product development, and societal growth.

Finally, PhD holders are highly skilled and knowledgeable individuals who go well beyond the realm of academia. PhD graduates will eventually possess a distinctive set of skills that are highly valued in non-academic professions because of their demanding training and intellectual hobbies. The wide skill set that PhD holders bring to the table is crucial, including their mastery of critical thinking, problem-solving, research and analytical skills, project management, and the capacity for autonomous work.

Indeed, PhD holders are skilled critical thinkers who can assess complicated issues from a variety of angles. Usually, they have developed the ability to assess material, spot underlying assumptions, and create conclusions based on the evidence through their in-depth investigation and intellectual exploration. PhDs are able to approach problems in non-academic professions with a high level of intellectual rigour and inventiveness, thanks to their talent for critical thought. PhD holders bring a vital viewpoint to their organisations by frequently challenging preconceived notions, posing novel theories, and offering creative solutions to hard issues. This encourages an environment that values growth and improvement. A PhD journey centres on problem-solving, as researchers take on cutting-edge and complex problems in their professions; therefore, PhD holders are skilled in decomposing difficult issues into manageable parts, carrying out methodical studies, and creating novel solutions. In positions outside of academia, their capacity to work through ambiguity, adjust to shifting conditions, and come up with original answers is crucial. PhD graduates provide an organised and analytical approach to problem-solving, helping them to solve complex difficulties with efficiency and effectiveness, whether it be refining company processes, developing new goods or services, or resolving operational challenges. Research and rigorous data analysis are hallmarks of PhD training. Research methodology, experimental design, data collection, and statistical analysis are areas in which PhD holders excel. PhDs' proficiency in compiling, analysing, and synthesising massive amounts of data enables them to produce insightful conclusions and promote data-driven decision-making. These abilities enable PhD holders to study market trends, spot patterns, assess risks, and direct strategic planning in non-academic settings. Their expertise in research and analysis enables firms to make wise decisions, streamline procedures, and remain ahead of the curve in a world that is becoming more and more data-centric.

In conclusion, a PhD requires excellent project management abilities. PhD holders are skilled in defining research goals, developing thorough work schedules, managing resources, and sticking to deadlines. Their expertise in managing complicated projects involving multidisciplinary teams and stakeholders comes from experience. Their capacity to drive ideas, carry out strategies, and guarantee project success benefits non-academic groups. Indeed, PhD graduates bring a methodical and systematic approach that promotes collaboration, reduces risks, and produces outcomes quickly and effectively.

PhDs wanting to apply for a non-academic job must strategically deal with the job market if they are to effectively transition into these positions. Accessing non-academic job options requires networking. PhD candidates and recent grads should make an effort to network professionally in their areas of interest. By going to conferences, seminars, and industry events, they can network with experts, share ideas, and learn about careers outside academia. LinkedIn[4] and other online networking sites offer ways to get in touch with potential employers and industry leaders. Additionally, enlisting the help of mentors in non-academic fields can be a great resource for advice and assistance during the job search. Finding fulfilling non-academic roles is made more likely by a strong professional network, which can provide access to untapped job prospects, referrals, and mentorship.

PhD candidates and alumni have a broad range of skills that are adaptable to jobs outside academia. Although having specific knowledge is necessary, employability can be considerably increased by learning skills like project management,

---

[4] http://linkedin.com/

communication, leadership, and data analysis. PhD students can take advantage of their coursework to practise managing collaborations, finances, and timetables. Teamwork and leadership abilities can be developed through involvement in interdisciplinary projects, student organisations, and volunteer work. Additionally, improving communication skills through education, conference presentations, or science communication platforms aids in bridging the gap between academia and industry. The desirability of PhD graduates to potential non-academic employment is increased by actively exploring opportunities to build transferrable abilities.

During their education as PhD candidates, PhDs should proactively look into internships or partnerships with businesses. Participating in non-academic organisations exposes one to a variety of workplaces, business processes, and professional networks. Look for chances to work on cooperative projects, conduct research together, or intern with business partners. These opportunities provide a first-hand understanding of the non-academic landscape, enabling PhD students to develop transferable abilities, forge connections with professionals, and make wise career selections. Industry partnerships can offer chances to apply research expertise to actual issues, thereby enhancing the value of a PhD in non-academic settings.

PhD candidates and recent graduates must be proficient communicators in order to be successful in the non-academic employment market. It's vital to emphasise transferrable talents, project management expertise, and problem-solving ability in resumes, cover letters, and interview responses. Place a strong emphasis on the results, effects, and accomplishments of research programmes that meet market demands. Translate complex concepts into comprehensible language to demonstrate the ability to communicate effectively across diverse audiences. Consider developing a professional online presence as well, such as a personal website or portfolio, to highlight your research achievements and abilities for positions outside of academia. PhD graduates can effectively communicate their value proposition to prospective jobs outside of academia by effectively translating their research experience.

A good reference is the article by Amy J. Ko titled 'Most Ph.D.s aren't professors',[5] which covers the various employment options open to those with a PhD. The author describes her own journey of learning about the many alternatives to the conventional path of becoming a tenure-track professor. According to Ko, many PhD students have the same goal of becoming professors when they first enrol in graduate school. However, the wide range of job prospects that a PhD can lead to is frequently kept a secret from undergraduates and even doctoral students. The National Science Foundation (NSF) estimates that fewer than 20% of PhD graduates work in academia.

---

[5] https://medium.com/bits-and-behavior/most-ph-d-s-arent-professors-13a741ef6868

Table 5.1 Career paths for PhD graduates include academia, industry, government, and entrepreneurship, from research and teaching to policy, consulting, and innovation

Position	Description
Tenure-track professors at prestigious research universities	These positions provide independence of thought and access to cutting-edge study.
Tenure-track professors at second-tier research universities	These jobs open up the classroom to more students from different backgrounds, allowing for a stronger emphasis on mentoring and teaching while still allowing for a significant research focus.
Soft-money professorships	More prevalent in the medical and health sciences, these jobs largely focus on research with little to no teaching responsibilities and include funding salaries through grants.
Post-docs	Doctoral students frequently obtain positions in other labs to carry out research under the guidance of professors, which can help them establish a record of published work.
Research scientists and research staff	These jobs entail carrying out research individually or as helpful team members of a research team, on large projects or in academia, as appropriate.
Tenure-track professors at top-tier teaching colleges	In these positions, excellence in teaching and mentoring is prioritised, with an emphasis on building relationships with students and encouraging student exploration.
Lecturers in research universities and teaching universities	These roles may not have expectations for research but have the chance to influence undergraduate culture and offer introductory courses.
Teachers in K–12 and community colleges	They play crucial roles that are frequently underappreciated, especially when working with diverse student groups.
Academic administration	Many PhD holders go on to assume important positions in schools and universities as managers of academic programmes, diversity initiatives, student recruitment, and other related tasks.
Industry research labs and industry R&D	These jobs centre on translating academic findings into real-world applications, whether it's forming new products and services or pursuing novel ideas.

Position	Description
User experience and data science	People with PhDs in a variety of disciplines can offer their research knowledge to firms to assist them in exploring product prospects and guiding decision-making.
Engineering	With a direct impact on the market, this job path involves delivering and maintaining products based on well-established research findings.
Management consultants	PhD competence and the capacity to make difficult ideas understandable are important in offering organisations strategic counsel.
Entrepreneurship	Some PhD holders join early-stage firms or launch their own businesses to commercialise their knowledge and findings.
Nonprofit executive	PhD holders can play important positions in non-profit organisations that are related to their field of specialisation and concentrate on resolving certain social issues.
Technology transfer contractors	PhDs can work as contractors on projects that bridge the gap between fundamental research and product design by identifying useful applications for fundamental discoveries.
Policy research	By working for organisations or think tanks performing contract research to guide policy decisions, PhDs can help shape laws and regulations.
Science funding and policy	To influence investments in science and comprehend the relationship between discovery and society, academics can pursue positions in national funding or policy organisations (such as the National Science Foundation or the European Research Council).
Government laboratories	Many government agencies, including the National Institutes of Standards and Technology or the National Security Agency, require PhD-level skills and offer opportunities to help the nation.
Science writing	PhD holders can use their knowledge to inform a wider audience outside of academic settings by writing about new scientific discoveries in popular books, science journalism, or as media experts.
Expert witness	PhD holders can testify as experts in court matters, explaining difficult concepts to juries and judges and frequently receiving payment for their work.

The author then goes on to outline and discuss many job options for PhD holders, as shown in the following Table 5.1.

The author underlines that each of these professional pathways is significant to society and that doctorate students must be aware of the range of options available outside of academia. By being aware of the variety of options available, students can make wise judgments and pursue career paths that are in line with their aspirations.

The PhD of the future

A PhD has traditionally been associated with becoming an expert in a particular topic, with a concentration on producing new knowledge within a certain discipline. The modern PhD is envisioned as a 'knowledge broker' rather than just a 'knowledge holder' who excels in their specialised field (Lazurko et al., 2020). This position of knowledge broker transcends boundaries; it makes linkages across academic fields, between various knowledge systems, and, crucially, between the scientific community and society at large.

A knowledge broker PhD understands that multidisciplinary approaches are frequently necessary to solve complicated, real-world challenges. They serve as links, bringing specialists from different domains to work together on complex problems. This guarantees that research findings have a more immediate and tangible influence on society while also expanding the breadth and originality of the study.

In addition, the knowledge broker PhD interacts with stakeholders outside academia, including the general public, business professionals, and legislators. They successfully convey study findings by converting difficult scientific ideas into language that is accessible to a wider audience, for example the Conversation[6] and other similar initiatives. The Conversation, a collaborative platform between academics and journalists, has emerged as one of the most prominent publishers of research-based news and analysis, producing articles crafted by professionals in tandem with digital technology experts, offering free access to their content with a mission to provide high-quality explanatory journalism for improved decision-making. This strategy makes sure that research tackles the most important problems of the day and contributes to the public's growing trust in science.

The future of education for aspiring PhD students must aim to equip them with a comprehensive understanding of transdisciplinarity. It must highlight how multifaceted these qualities are, spanning a diverse array of skills and encompassing

---

[6] http://theconversation.com/

a wide range of abilities, character traits, and a specific way of thinking that is in line with a transdisciplinary outlook.

Such a curriculum recognises the need for more research into the subjective and embodied experiences of people moving into transdisciplinary roles, for instance, by incorporating a transdisciplinary framework (Augsburg, 2014).

It must emphasise the importance of conducting studies on the individual paths, difficulties, and life-changing events that PhD candidates encounter as they develop into transdisciplinary researchers. The programme's goals will be to further students' understanding of the human elements involved in the process of becoming skilled transdisciplinary individuals by supporting these kinds of research activities.

As discussed, the conventional idea of a PhD title is about to change in the dynamic environment of academia and industry. It is anticipated that subsequent versions of the PhD would cover a wider range of abilities and proficiencies than merely in-depth subject matter knowledge. This shift could entail putting more of a focus on multidisciplinary education, encouraging flexibility and creativity, and incorporating practical applications into research projects. Furthermore, future PhD programmes might include experiential learning components that encourage applicants to pursue public policy efforts, industrial partnerships, or entrepreneurial endeavours in addition to their academic goals. Additionally, there may be a shift in the appreciation of various forms of scholarship, recognising contributions such as open science, community involvement, and the creation of useful technology or solutions.

In PhD education, experiential learning will be a shift from traditional classroom-based instruction to a more hands-on and interactive methodology. It can take many different forms, including industrial placements, cooperative projects with external partners, community engagement programmes, internships, and entrepreneurial experiences. Through these experiences, doctorate candidates can put their theoretical knowledge to use in practical situations and acquire priceless insights, abilities, and viewpoints that they might not otherwise be able to obtain within the walls of academia. Additionally, through community engagement programmes, PhD candidates can work with stakeholders or local communities to adapt their findings to specific community needs or societal concerns. This engagement fosters a sense of social responsibility and the capacity to explain difficult concepts to a variety of audiences in addition to demonstrating the relevance of their study.

Contemporary trends within doctoral programmes increasingly advocate for students to explore facets within their research endeavours that exhibit potential for commercialisation. They could gain knowledge about start-up formation, intellectual property, and the process of bringing an idea from the lab to the market. This helps them develop an entrepreneurial attitude and gives them the tools they need to negotiate the nexus between academia and business successfully.

All things considered, adding experiential learning to PhD programmes enhances the educational process and gives graduates a more varied skill set, flexibility, and a clearer sense of how their work might be applied outside of academic settings. By representing a near-future of PhD education, it also equips them with the skills necessary to successfully negotiate the challenges of a quickly evolving labour market and make significant contributions to society in a variety of ways.

Summary

This chapter offered advice on planning for the future, from pursuing a post-doctoral position in academia, industry, or government to exploring other career paths such as science communication or entrepreneurship. Additionally, guidance on writing grant proposals is offered to secure funding for research, which requires strong writing and project management skills. For those interested in pursuing an academic career, there was guidance on running a research lab, mentoring students, and building a research programme. This chapter concluded by addressing the recent debate around the future of the PhD and considering how it might change.

Epilogue

Well done! You've reached the conclusion of a book that aims to offer thorough advice on conducting research and effectively earning a PhD. We have looked at many different facets of the research process throughout the chapters, including the intricate details of selecting a methodology, formulating a research topic, analysing data, writing academically, and much more.

Whether you are an early career academic just beginning your teaching and mentoring position or a recent PhD graduate beginning to supervise students, navigating the responsibilities and dynamics of mentorship can be both exciting and challenging. This book has offered advice on a range of topics, including managing student expectations, giving helpful criticism, negotiating the challenges of supervising PhD students, and coping with viva preparation.

If you are a PhD candidate, then the path to earning a doctorate is a wonderful one, replete with struggles, successes, and opportunities for personal development. You have shown a dedication to advancing the boundaries of knowledge and furthering your area by choosing this path. The goal of this book was to give you the skills, knowledge, and useful guidance you need to successfully complete this challenging journey. I have made an effort to empower you with the knowledge and abilities required to succeed in your research pursuits, drawing on both my personal experiences and the body of available literature.

From the initial stages of conducting a thorough literature review to the final steps of writing and presenting your research findings, we have explored each phase of the research process in detail. We discussed the importance of formulating a research question that addresses a unique contribution to your field, and we examined different research methodologies, theoretical frameworks, and conceptual frameworks that can guide your investigations. Additionally, we explored the significance of effective knowledge organisation, planning, and data analysis, ensuring that your research is both rigorous and insightful.

To support your research journey, we also delved into essential tools such as reference management software and AI-based tools, which can streamline your work and enhance your productivity. Furthermore, we discussed the construction of your own PhD toolkit, highlighting the importance of knowledge organisation and sharing valuable resources and tips to facilitate your research process.

Receiving feedback and engaging in academic conferences and events were also key topics of this book. The valuable insights gained from expert feedback can help refine your research, enhance your academic writing, and strengthen the impact of your work. Similarly, participating in conferences and events provides opportunities for dissemination, networking, and collaboration, enabling you to establish yourself as a respected voice in your field.

The final chapter of this book examines the opportunities that lie ahead when you progress beyond earning your doctorate. We have taken into account options like drafting grant bids and the possibility of starting and managing your own research lab. These skills give you the chance to advance your contributions to your profession, influence the course of research, and mentor younger generations.

To help you on your research journey, I have made an effort to offer helpful tips, specific examples, and insightful commentary throughout this book. I hope the chapters have made important ideas and techniques clearer, enabling you to study and review the basics of each subject.

A student's guidance in putting together a unique PhD toolkit is given in addition to the resources offered in this book. Along with your research, you can use the additional tools and advice in this compilation to help you overcome any obstacles that may come up.

Lastly, I've added a special feature for supervisors in their early careers: grey boxes. These observations and suggestions will offer assistance to academics who are starting their careers as PhD student supervisors and will offer insightful advice based on the material covered in this book.

PhD grads should keep in mind that their journey does not finish when they earn their doctorate, as they begin the next phase of their career. Instead, it heralds the start of a brand-new stage brimming with promising opportunities. Accept the abilities, information, and self-assurance you have acquired and keep pursuing your enthusiasm for investigation, invention, and discovery.

As you start your research journey, may this book serve as a guiding light, providing ideas, suggestions, and insights to help you overcome the obstacles on your path. Keep in mind that your contributions could influence your field's future and have a long-lasting effect on society.

As the pages of this journey come to a close, I must acknowledge the significant influence that mentorship has had on my life. Isaac Newton once said, 'If I have seen further, it is by standing on the shoulders of giants.' Indeed, the knowledge, direction, and inspiration passed down to us by those who came before us have helped us to advance our level of awareness and knowledge. True mentorship, on the other hand, goes beyond imitation and focuses on creating a safe space for development and self-awareness. According to Steven Spielberg's insightful observation, 'The delicate balance of mentoring someone is not creating them in your own image, but giving them the opportunity to create themselves.' You can take the lead, carrying the torch of your mentors with appreciation, eager to become a mentor yourself and enable others to forge their own unique paths of brilliance and innovation.

Bibliography

Alreck, P. L., & Settle, R. B. (1995). The survey research handbook. McGraw-Hill/Irwin.

Alvesson, M., & Sandberg, J. (2013). Constructing research questions: Doing interesting research. Sage.

Anderson, C. (2013). How to give a killer presentation. Harvard Business Review, 91(6), 121–125.

Anfara, Jr., Vincent, A., & Mertz, N. T. (2014). Theoretical frameworks in qualitative research. Sage publications.

Atkinson, P. (2007). Ethnography: Principles in practice. Routledge.

Augsburg, T. (2014). Becoming transdisciplinary: The emergence of the transdisciplinary individual. World Futures, 70(3–4), 233–247.

Avison, D. E., Shaik, Z., Malaurent, J., Gaur, A., & Mousavi, R. (2013). The supervisor-student relationship: The problem of conflicting 'mixed metaphors'. In Bled eConference (p. 2).

Braunschweig, K., Eberius, J., Thiele, M., & Lehner, W. (2012). The state of open data. Limits of Current Open Data Platforms, Conference Proceedings World Wide Web Conference, WWW 2012, Lyon.

Bernoff, J. (2016). Writing without bullshit: Boost your career by saying what you mean. HarperCollins.

Birks, M., & Mills, J. (2015). Grounded theory: A practical guide. Sage.

Blanchard, K., & Johnson, S. (1983). The one-minute manager. Cornell Hotel and Restaurant Administration Quarterly, 23(4), 39–41.

Bloudoff-Indelicato, M. (2015). NIH metric that assesses article impact stirs debate. Available at: https://www.nature.com/news/nih-metric-that-assesses-article-impact-stirs-debate-1.18734 (accessed 1 December 2020).

Bryson, B. (2004). A short history of nearly everything. Broadway.

Card, A. J. (2017). The problem with '5 whys'. BMJ Quality and Safety, 26(8), 671–677.

Campbell, D. T., & Stanley, J. C. (2015). Experimental and quasi-experimental designs for research. Ravenio Books.

Cassuto, L. (2018). On the dissertation: How to write the introduction. The Chronicle of Higher Education. https://www.chronicle.com/article/On-the-Dissertation-How-to/243507

Czocher, J. A. (2014). Toward building a theory of mathematical modelling. North American Chapter of the International Group for the Psychology of Mathematics Education.

De Bono, E., & Zimbalist, E. (1970). Lateral thinking (pp. 1–32). Penguin.

De Lara, J. (2020, June 11th). Scientific writing. https://www.slideshare.net/slideshow/scientific-writing-235422422/235422422

Dix, A. (2004, February 8th). Research and innovation techniques. https://www.alandix.com/academic/topics/res-tech/

Dix, A., Ormerod, T., Twidale, M., Sas, C., Silva, P. A., & McKnight, L. (2006). Why bad ideas are a good idea. In Proceedings of HCIEd.2006-1 Inventivity, Ballina/Killaloe. http://www.hcibook.com/alan/papers/HCIed2006-badideas/

Dunleavy, P. (2014). Why do academics choose useless titles for articles and chapters? Four steps to getting a better title. Impact of Social Sciences Blog.

Edwards, P. N. (1998). How to give an academic talk: Changing the culture of public speaking in the humanities. Accessible online at http://www.si.umich.edu/pne/PDF/howtotalk.pdf. Copyright, 2004.

Elmqvist, N. (2016, 19th November). Writing Rebuttals. https://niklaselmqvist.medium.com/writing-rebuttals-7f6949eddf6e

Eppler, M. J. (2006). A comparison between concept maps, mind maps, conceptual diagrams, and visual metaphors as complementary tools for knowledge construction and sharing. Information Visualization, 5(3), 202–210.

EUR Library – Erasmus University Rotterdam. (n.d.). *Home* EUR Library. YouTube. Retrieved January 10, 2025. From https://www.youtube.com/@LibraryEUR

Franklin University. What is a doctorate: Everything you need to know. https://www.franklin.edu/blog/what-is-a-doctorate-degree

Gewin, V. (2018). How to write a first-class paper. Nature, 155, 129–130.

Gladfelter, A. S., & Peifer, M. (2017). What your PI forgot to tell you: Why you actually might want a job running a research lab. Molecular Biology of the Cell, 28(13), 1724–1727.

Guo, P. (2012, November). Advice for early-stage PhD students. https://cacm.acm.org/blogcacm/ph-d-s-from-the-facultys-perspective/

Hahanov, V., Hacimahmud, A. V., Litvinova, E., Chumachenko, S., & Hahanova, I. (2018, September). Quantum deductive simulation for logic functions. In 2018 IEEE East-West Design & Test Symposium (EWDTS) (pp. 1–7). IEEE.

Harzing, A-W. (2023). Publish or perish. https://harzing.com/resources/publish-or-perish – Sat 6 Feb 2016 16:10 (updated Fri 25 Aug 2023 16:24).

Heath, C., & Heath, D. (2007). Made to stick: Why some ideas survive and others die. Random House.

Hevner, A., & Chatterjee, S. (2010). Design science research in information systems. In Design research in information systems (pp. 9–22). Springer.

Hong, J. (2013). Ph. D. students must break away from undergraduate mentality. Communications of the ACM, 56(7), 10–11.

Hu, F., He, F., & Yaron, D. J. (2023). Treating semiempirical hamiltonians as flexible machine learning models yields accurate and interpretable results. Journal of Chemical Theory and Computation, 19(18), 6185–6196.

Jaksch, M. (2023). 67 Top tools for writers and bloggers in 2023. https://writetodone.com/top-writing-blogging-tools/

Jannach, D., Zanker, M., Felfernig, A., & Friedrich, G. (2010). Recommender systems: An introduction. Cambridge University Press.

Kuhn, T. S. (2012). The structure of scientific revolutions. University of Chicago Press.

Laubepin, F. (2013). Interuniversity consortium for political and social research. https://tinyurl.com/y2osrhdp

Langham-Putrow, A., Bakker, C., & Riegelman, A. (2021). Is the open access citation advantage real? A systematic review of the citation of open access and subscription-based articles. PloS One, 16(6), e0253129.

Laymon, R. (1989). Applying idealized scientific theories to engineering. Synthese, 81, 353–371.

Lazar, J., Feng, J. H., & Hochheiser, H. (2017). Research methods in human-computer interaction. Morgan Kaufmann.

Lazurko, A., Alamenciak, T., Hill, L. S., Muhl, E. K., Osei, A. K., Pomezanski, D., & Sharmin, D. F. (2020). What will a PhD look like in the future? Perspectives on emerging trends in sustainability doctoral programs in a time of disruption. World Futures Review, 12(4), 369–384.

Lee, A., & Bongaardt, R. (Eds.) (2021). The future of doctoral research: Challenges and opportunities. Routledge.

Lehmann, E. L. (2011). Fisher, Neyman, and the creation of classical statistics. Springer Science & Business Media.

Medium (2015, March 21st). Writing for Research. Assessing your research and publication choices. https://tinyurl.com/y7ugfjyx

Meyer, B. (2016, March 3rd). What's your research? https://cacm.acm.org/blogcacm/whats-your-research/

Miedema, F. (2022). Open science: The very idea. Springer Nature.

Moher, D., Liberati, A., Tetzlaff, J., Altman, D. G., & Prisma Group. (2009). Preferred reporting items for systematic reviews and meta-analyses: The PRISMA statement. PLoS Med, 6(7), e1000097.

Nacke, L. E. (2017, May). How to write and review Chi papers. In Proceedings of the 2017 Chi Conference Extended Abstracts on Human Factors in Computing Systems (pp. 1228–1231).

Narayanan, A. (2018, 15th May). How to constructively review a research paper. https://freedom-to-tinker.com/2018/05/15/how-to-constructively-review-a-research-paper/

Neyman, J., & Pearson, E. S. (1933, October). The testing of statistical hypotheses in relation to probabilities a priori. In Mathematical proceedings of the Cambridge philosophical society (Vol. 29, No. 4, pp. 492–510). Cambridge University Press.

Piwowar, H., Priem, J., Larivière, V., Alperin, J. P., Matthias, L., Norlander, B., & Haustein, S. (2018). The state of OA: A large-scale analysis of the prevalence and impact of Open Access articles. PeerJ, 6, e4375.

Rachit, N. (2019, 3rd March). Project management for PhD students. https://rachit.pl/post/project-management/

Novak, J. D., & Cañas, A. J. (2008). The theory underlying concept maps and how to construct and use them. Florida Institute for Human and Machine Cognition, 1(1), 1–31.

Pauk, W., & Owens, R. J. (2010). How to study in college (Chapter, 10, pp. 235–277). Wadsworth.

Pinker, S. (2015). The sense of style: The thinking person's guide to writing in the 21st century. Penguin Books.

Privitera, G. J. (2016). Statistics for the Behavioural Science. Sage Publications.

Tahan, Rana. (2022). Typical research article structure. Source: https://www.researchgate.net/publication/359815719_Typical_Research_Article_Structure

Ravitch, S. M., & Riggan, M. (2012). Reason & rigor: How conceptual frameworks guide research. Sage.

Ringmar, E. (2015). How to write an academic paper. Department of Political Science, Lund University.

Rockinson-Szapkiw, A. J., & Spaulding, L. S. (2014). Navigating the doctoral journey: A handbook of strategies for success. Rowman & Littlefield.

Shor, P. W. (1994, November). Algorithms for quantum computation: Discrete logarithms and factoring. In Proceedings 35th annual symposium on foundations of computer science (pp. 124–134). Ieee.

Simons, H. (2009). Case study research in practice. Sage Publications.

Stewart, J. J. P. (1989). Optimization of parameters for semiempirical methods I. Method Journal Computational. Chemistry, 10, 209–220.

Stinger, E. T. (1996). Action research: A handbook for practitioners. SAGE Publications, Inc.

Stone, G. (2015, 12th November). For Better Presentations, Start with a Villain. https://hbr.org/2015/11/for-better-presentations-start-with-a-villain

Sword, H. (2017). Air & light & time & space: How successful academics write. Harvard University Press.

Tay, A. (2020). How to write a superb literature review. Nature. https://doi.org/10.1038/d41586-020-03422-x

Teperek, M. (2018). How to make the most of an academic conference–a checklist for before, during and after the meeting". Impact of Social Sciences blog, 16.

Van de Ven, A. H. (2007). Engaged scholarship: A guide for organizational and social research. Oxford University Press on Demand.

Vohland, K., Land-Zandstra, A., Ceccaroni, L., Lemmens, R., Perelló, J., Ponti, M., & Wagenknecht, K. (2021). The science of citizen science (p. 529). Springer Nature.

Watson, M. (2016, 27th October). Tips for PhD students and early-career researchers. http://www.opiniomics.org/tips-for-phd-students-and-early-career-researchers/

Watson, P. (2006). Ideas: A history from facts to freud. Weidenfeld & Nicolson.

Wheelan, C. (2013). Naked statistics: Stripping the dread from the data. WW Norton & Company.

Wohlin, C., Runeson, P., Höst, M., Ohlsson, M. C., Regnell, B., & Wesslén, A. (2012). Experimentation in software engineering. Springer Science & Business Media.

Yu, M. C., Wu, Y. C. J., Alhalabi, W., Kao, H. Y., & Wu, W. H. (2016). ResearchGate: An effective altmetric indicator for active researchers?. Computers in Human Behavior, 55, 1001–1006.

Zhu, X., Turney, P., Lemire, D., & Vellino, A. (2015). Measuring academic influence: Not all citations are equal. Journal of the Association for Information Science and Technology, 66(2), 408–427.

Zobel, J. (2004). Writing for computer science (Vol. 8). Springer.

Index

2 04